Natural Stone and World Heritage

Natural Stone and World Heritage
Series editor: Dolores Pereira

Natural Stone and World Heritage
Salamanca (Spain)
Dolores Pereira

Natural Stone and World Heritage
Delhi-Agra, India
Gurmeet Kaur, Sakoon N. Singh, Anuvinder Ahuja and Noor Dasmesh Singh

Natural Stone and World Heritage
UNESCO Sites in Germany
Edited by Angela Ehling, Friedrich Häfner and Heiner Siedel

For more information about this series, please visit: www.routledge.com/Natural-Stone-and-World-Heritage/book-series/NSWH

Natural Stone and World Heritage

UNESCO Sites in Germany

Edited by

Angela Ehling
BUNDESANSTALT FÜR GEOWISSENSCHAFTEN UND
ROHSTOFFE (BGR), BERLIN, GERMANY

Friedrich Häfner
BUDENHEIM, GERMANY

Heiner Siedel
TU DRESDEN, INSTITUTE OF GEOTECHNICAL ENGINEERING,
DRESDEN, GERMANY

CRC Press
Taylor & Francis Group
Boca Raton London New York

CRC Press is an imprint of the
Taylor & Francis Group, an **informa** business

Cover image: Statues of the founders Uta and Ekkehard made of limestone (Freyburger Schaumkalk) from the west choir of the Naumburg Cathedral (around 1250)

CRC Press/Balkema is an imprint of the Taylor & Francis Group, an informa business

Typeset by Apex CoVantage, LLC

Library of Congress Cataloging-in-Publication Data
Names: Ehling, Angela, editor. | Häfner, Friedrich, editor. | Siedel, Heiner, editor.
Title: Natural stone and world heritage : UNESCO sites in Germany / edited by: Angela Ehling, BGR, Berlin, Germany, Friedrich Häfner, Budenheim, Germany, Heiner Siedel, Institute of Geotechnial Engineering, TU Dresden, Dresden, Germany.
Description: Boca Raton : CRC Press, [2022] | Includes bibliographical references and index.
Identifiers: LCCN 2021012179 (print) | LCCN 2021012180 (ebook)
Subjects: LCSH: Building stones—Germany—History. | Historic buildings—Germany—Materials. | Building, Stone—Germany—History. | Geology—Germany.
Classification: LCC TN953.G3 N38 2022 (print) | LCC TN953.G3 (ebook) | DDC 691/.20943—dc23
LC record available at https://lccn.loc.gov/2021012179
LC ebook record available at https://lccn.loc.gov/2021012180

Published by: CRC Press/Balkema
            Schipholweg 107C, 2316 XC Leiden, The Netherlands
            e-mail: Pub.NL@taylorandfrancis.com
            www.routledge.com – www.taylorandfrancis.com

ISBN: 978-0-367-42260-8 (hbk)
ISBN: 978-0-367-82306-1 (eBook)

DOI: https://doi.org/10.1201/9780367823061

Contents

UNESCO World Heritage Sites in Germany and natural stones

Introduction

Germany, in the middle of Europe, reflects European history, which is often "written" in stone. The oldest European stone witnesses are menhirs and megalithic buildings, which are known all over the world. Consisting of stones that could be found nearby, they give evidence of life during the Stone Age. Since Roman times, stone buildings and constructions at high technical level appeared also north of the Alps, including the territory of Germany. The technology and knowledge to process natural stones and to use them for construction purposes came along with the territorial expansion of the Roman Empire. The further evolution of stone construction was interrupted by a period of some hundred years after the fall of the Roman Empire, when stone buildings were small, simple and rare. Since Carolingian times, there is again evidence of important stone buildings in Germany. From that time, natural stone as construction material has been closely accompanying the development of different architectural styles. Regional stone materials were used for the construction of huge cathedrals, monasteries, churches and excellent sculptures in the Gothic period. Town halls and living houses dating from the Renaissance period often comprise architectural elements made of natural stone as well. In baroque time, splendid palaces and churches were built of stone, the interior of which was decorated with altars, columns, staircases and floors made of polished marbles and limestones. Parks and gardens of the same period contain stone sculptures and smaller stone buildings. Natural stone has been one of the most important materials for massive constructions until the beginning of the 20th century. Even on modern buildings, natural stones are used for decoration and façade cladding, although they are constructed of concrete, steel and glass today.

The World Heritage Committee of the UNESCO has designated more than 46 Natural and Cultural World Heritage Sites in Germany till this day. Among them are different kinds of buildings dating from the last

2,000 years. At least at 30 of these Cultural Heritage Sites, natural stones are apparent and substantial for the image of the site (Figure 1). In most cases, they reflect the surrounding natural countryside as well as social, political and economic conditions at the time of construction and during their lifetime. Although these sites and objects are extensively presented in historical, architectural and cultural context, the historic construction material is ignored in most cases. However, the natural stones with their specific color, texture and appearance give many of these sites their distinctive face.

The book focuses on traditional building stones used as construction material for the UNESCO World Heritage Sites and their buildings in Germany. It is due to the limited volume of a book of this series as well as to the ongoing process of UNESCO World Heritage designation that it cannot be exhaustive. Ten selected German Heritage Sites are exemplarily presented in more detail with their history and their building and decorative stones (Figure 1). Moreover, comments on the weathering state of the respective stone materials and restoration measures are made.

In a second part, important stones, which had been used at German World Heritage Sites, are presented with their occurrences, aspects of quarrying in historical times as well as their petrographical, mineralogical and technical features, including weathering behavior.

The idea is to demonstrate the strong connection of historic building stones with the geology of the surrounding countryside as well as with possibilities of quarrying, transport and processing.

Since the subject comprises detailed information about building stones, geology and architectural history from different regions of Germany, several geologists and conservation scientists who are very familiar with certain buildings and building stones contributed to this book.

The list of references shall provide opportunity to interested readers to deepen their knowledge about special building stone and conservation issues.

The editors hope that this book might contribute to make the rich diversity of historic building stones in Germany more popular.

December 2020
Angela Ehling, Friedrich Häfner and Heiner Siedel

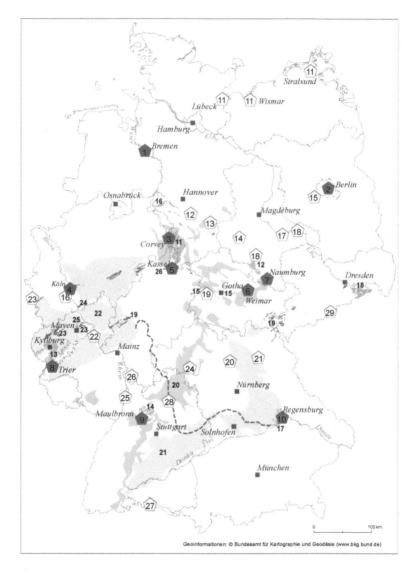

Figure I

UNESCO Cultural World Heritage Sites in Germany made with natural stones

① 1 Bremen: Town Hall and Roland on the Marketplace
② 2 Berlin Museum Island
③ 3 Corvey: Carolingian Westwork and Civitas
④ 4 Cologne Cathedral
⑤ 5 Kassel: Bergpark Wilhelmshöhe
⑥ 6 Classical Weimar and Bauhaus in Weimar
⑦ 7 Naumburg Cathedral
⑧ 8 Trier - Roman and medieval World Heritage Sites
⑨ 9 Maulbronn Monastery Complex
⑩ 10 Regensburg: Old Town
⑪ 11 Hanseatic cities: Lübeck, Wismar, Stralsund
⑫ 12 Hildesheim: St Mary's Cathedral and St Michael's Church
⑬ 13 Goslar: Historic Town and Mines of Rammelsberg
⑭ 14 Quedlinburg: Collegiate Church, Castle, and Old Town
⑮ 15 Potsdam and Berlin: Palaces and Parks
⑯ 16 Brühl: Castles of Augustusburg and Falkenlust
⑰ 17 Dessau-Wörlitz: Garden Kingdom
⑱ 18 Eisleben and Wittenberg: Luther memorials
⑲ 19 Wartburg Eisenach
⑳ 20 Bamberg Town
㉑ 21 Bayreuth: Margravial Opera House
㉒ 22 Upper Middle Rhine Valley
㉓ 23 Aachen Cathedral
㉔ 24 Würzburg: Residence
㉕ 25 Speyer Cathedral
㉖ 26 Lorsch: Abbey and Altenmünster
㉗ 27 Monastic Island of Reichenau
㉘ 28 Frontiers of the Roman Empire
㉙ 29 Mining Cultural Landscape Erzgebirge/Krušne hoří

⬟ presented in this book
⬠ not presented

Natural Stones at these UNESCO sites

11	Weser Sandstone
12	Nebra Sandstone
13	Kylltal Sandstone
14	Schilfsandstein SW-Germany
15	Thuringian Rhaetian Sandstones
16	Obernkirchen Sandstone
17	Regensburg Green Sandstone
18	Elbe Sandstone
19	Coloured "Marbles"
20	Muschelkalk Germany
21	Jurassic Limestones
22	Roof Slates
23	Rhenish Basaltlava
24	Drachenfels Trachyte
25	Rhenish Tuff
26	Habichtswald Tuff

numbers = chapter numbers
(not in every coloured area)

Chapter I

UNESCO sites in Germany

Contents

1.1 Town Hall and Roland on the marketplace of Bremen

Jürgen Pätzold

Brief description

The Town Hall and Roland on the marketplace of Bremen in northwest Germany are recognized as an outstanding ensemble, representing the civic autonomy and sovereignty that developed in the Holy Roman Empire in Europe (UNESCO WHC 2004; Elmshäuser et al. 2002, Figure 1.1.1). The Old Town Hall was constructed in the Gothic style at the close of the Middle Ages in the early 15th century. The stone statue of Roland, erected in 1404, stands in front of the town hall, symbolizing the rights and privileges of the free and imperial city of Bremen. The medieval building was renovated and modified in the style of the so-called Weser Renaissance during the early 17th century. The New Town Hall was built adjacent to the old one in the early 20th century in Neo-Renaissance style. The complete ensemble escaped destruction during World War II and was included in the list of UNESCO World Heritage Sites in 2004.

Figure 1.1.1 Bremen Town Hall, southwest façade as seen from the marketplace.

Old Town Hall

The Old Town Hall was built as a Gothic hall structure in the years 1405–1410. It is a transverse rectangular two-story hall building (*Saalge-schossbau*), a representative of the palas type, with an arcade extending along its entire width and a balustrade beneath a high roof. The façade of the Gothic town hall is characterized by the combination of alternating layers of raw and black-glazed bricks, with the extensive use of ashlar. For lack of natural stone resources in the North German lowlands, ashlars were transported from the catchment area of the Weser River and its tributaries. Sandstones of Mesozoic age were primarily used. The procurement of building materials from the area around Hannover is alluded to in a number of entries in the preserved book of accounts related to the town hall construction. Fine homogeneous grey sandstones (called *"Graustein"* or *"Grauwerk"*) from the Wealden facies of the Early Cretaceous (see Chapter 2.6) in the Weser Mountains were used in large quantities. Massive Red Weser Sandstones from the Middle Buntsandstein (see Chapter 2.1) of the upper Weser region were used as floor tiles in the cellar (*Ratskeller*) and plates from laminated Red Weser Sandstone were utilized for roofing. Brown Porta Sandstone from the Middle Jurassic was only used in smaller quantities. In addition, Hannover limestones and mussel shells from the North Sea were brought in for local production of mortar. The Bremen account book also names a number of stonemasons and stone carvers. Sixteen stone figures by master sculptor Johan and his journeymen were placed between the large windows of the first upper story, all of them standing on consoles beneath canopies. The original Gothic statues of the emperor and seven electoral princes are preserved in the Focke Museum, Bremen, while copies from Obernkirchen Sandstone were placed on the main town hall façade in 1960–1961. The preserved figures on the northwest and southeast side were originally intended to represent prophets from the bible (Gramatzki 1994).

Bremen Roland

Even before construction of the old town hall began in 1405, a statue of Roland had been erected on the marketplace in Bremen (Figure 1.1.2). An entry in the Bremen book of accounts states that the council of Bremen had a Roland built of stone, which cost 170 Bremer Marks and was invoiced by Clawes Zeelsleghere and Jacob Olde.

The 5.5 m high stone statue of Roland was constructed using Elm limestone (*Elm-Kalkstein*) and placed on a stepped base with a supporting column topped by a Gothic-ornamented canopy. The monument reaches a total height of 10.2 m and is the first free-standing statue of the German Middle Ages. Historical representations document that it was painted with intense coloring during its early centuries. The limestone material came from quarries still known today from the Elm, south of Königslutter. The Bremen Roland has been repeatedly restored over the centuries and individual stones of Elm limestone (*Elm-Kalkstein*) have been replaced by Obernkirchen Sandstone. However, it can still be regarded as original because the replacement was performed stone by stone. The original head from 1404 was not replaced until 1983, and it is now in the museum. The copy was made of original material of the *Elm-Kalkstein*, which represents a shallow-marine facies of the Terebratel Beds of the Lower Muschelkalk. The copy exhibits the characteristic cross-bedding (Figure 1.1.3), while the original head from 1404 shows the typical grey weathering crusts on the surface. The last active quarry in the Elm limestone at Hainholz near Königslutter was abandoned around 2005 (Lepper et al. 2018). Today, the Bremen Roland is considered to be the oldest and largest statue of this type still in place.

Weser Renaissance

Under the direction of master builder Lüder von Bentheim, the Bremen town hall was extensively renovated and reconstructed at the beginning of the 17th century in the style of the so-called Weser Renaissance of

Figure 1.1.2 Roland on the marketplace.

Figure 1.1.3 Head of Roland.

Northern Germany and was richly decorated using Obernkirchen Sandstone after Dutch models (Figure 1.1.1). The use of Obernkirchen Sandstone in the Weser region dates back to the 11th century (see Chapter 2.6). Due to its very favorable material properties, the sandstone was increasingly used by stone sculptors with the spread of the Weser Renaissance. Both raw and prefabricated building blocks were transported by horse carts from the Bückeberge near Obernkirchen to the Weser near Rinteln, where they were loaded onto oak barges (so-called *Eken*) then taken downstream to Bremen to be stored and further processed (Kuster-Wendenburg 2002). Because of their light grey weathering color, they were known by the trade name "greystone". From Bremen, the sandstones were also later sold overseas and thus became known as *Bremer Stein*.

New Town Hall

Between 1909 and 1913, the Bremen town hall was extended in Neo-Renaissance style, according to the plans of the Munich architect Gabriel von Seidl (Figure 1.1.4). In the style of the Old Town Hall, the exterior façades were designed using a mixture of Oldenburg *Handstrich* clinker and Southern German Muschelkalk limestone (*Fränkischer Muschelkalk*). The extensive masonry work on the New Town Hall was carried out by the Bavarian sculptor Julius Seidler and Heinrich Erlewein of Bremen. Obernkirchen Sandstone and Untersberg "marble" were used for the festival stairway, and mosaic tiles of white marble as well as black and red nodular limestone decorate the upper lobby.

Figure 1.1.4 Old and New Town Hall.

1.2   Museum Island (*Museumsinsel*) Berlin

Gerda Schirrmeister

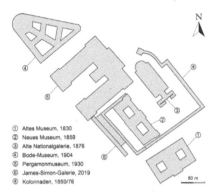

① Altes Museum, 1830
② Neues Museum, 1859
③ Alte Nationalgalerie, 1876
④ Bode-Museum, 1904
⑤ Pergamonmuseum, 1930
⑥ James-Simon-Galerie, 2019
④ Kolonnaden, 1859/76

Figure 1.2.1 Site plan of the Museum Island, Berlin.

Designed by J. Rätz.

Building history

The unique ensemble of five museums was built in the center of Berlin in the northern part of a Spree island between 1824 and 1930. The Museum Island Berlin is recognized as the most outstanding example of the concept of art museum owing its origin to the Age of Enlightenment in the 18th century (UNESCO WHC 1999). It was included in the list of UNESCO World Heritage Sites in 1999.

Access to the former princely collections for all people was claimed since the French Revolution in 1793. In 1810, the Prussian king Friedrich Wilhelm III commissioned Wilhelm von Humboldt to compile a well-selected art collection open to the public. In 1830, the first museum of this kind, later called Old Museum (*Altes Museum*), opened. In 1841, the Prussian king Friedrich Wilhelm IV issued an order for the further development of the northern Spree island to a sanctuary of art and science.

Some of the most renowned Prussian architects – Karl Friedrich Schinkel, Friedrich August Stüler, Ernst Eberhard von Ihne and Alfred Messel – designed the museums. The architecture illustrates the evolution of modern museum design; there are direct connections between the buildings and the collections. The representative façades were built with different natural stones, which reflect the general development of their use in the architectural and building history of Berlin in the course of about 100 years.

Elbe Sandstones (see Chapter 2.8) have been used in Berlin since the 15th century because these sandstones were the nearest significant occurrence of building stones, which could be transported on the river. They were used sparingly for important architectural elements of Old and New Museum (1830 and 1859): base, architectural jewelry and for the 18 colossal Ionic columns (Figure 1.2.2). Nebra Sandstone (see Chapter 2.2) was used in Berlin only during a relatively short period between 1846 and 1880. The Old National Gallery (*Alte Nationalgalerie*, 1876) was the first public building in Berlin completely faced with sandstone since the Brandenburg Gate (1788). Since 1871, Silesian Sandstones could be transported to the imperial capital by railway. The Bode Museum was colored with different varieties of these sandstones. The Pergamon Museum, the first architectural museum of the world, shows the typical limestone façades, which emerged in Berlin at the beginning of the 20th century.

The museums had been heavily damaged during World War II. They were initially repaired using mostly East German and Eastern European stones for replacement. The National Gallery reopened in 1949, the Bode Museum and the Pergamon Museum followed in the 1950s, the Old Museum in 1966. The New Museum remained in ruins.

In 1999, the Museum Island Master Plan was adopted with construction measures to preserve the monuments and meet the requirements of a contemporary museum complex. The Old National Gallery and the Bode Museum had been completely restored up to 2001–2006. The New Museum was very elaborately rebuilt and carefully restored up to 2009. A new entrance building was built for central service functions, the James Simon Gallery, which opened in 2019. The Pergamon Museum is currently under restoration, and the Old Museum has yet to be completely renovated.

Essential natural stone utilizations are assigned to the individual buildings discussed next. There is no claim to completeness (see also Schirrmeister 2006, 2009, 2013).

Old Museum (Altes Museum) with Granite Super Bowl

Architect: Karl Friedrich Schinkel (1781–1841)
Natural stones outside:

- Elbe Sandstones, especially from *Kirchleite* for the columns and flooring of the landing (partly repaired with material from *Reinhardtsdorf*) and *Posta* type for the base
- *Lausitzer Granit* (grey Cambrian granodiorite from Saxony) for the outdoor staircase

Figure 1.2.2 Main front of the Old Museum with 18 Ionic sandstone columns and the Granite Super Bowl in front of the Museum.

- Assuan Rose Granite (red Precambrian granite from Egypt) for two tubs from the 3rd century in the columned hall

Natural stones inside:

- *Cottaer Sandstein* (variety of Elbe Sandstone, see Chapter 2.8) for staircase and columns
- *Theumaer Fruchtschiefer* (dark grey slate, Ordovician slate from Saxony) in the center of the flooring in the rotunda
- *Großkunzendorfer Marmor* (light grey Precambrian marble from Silesia, Poland) for flooring for the ground floor
- Saalburg "Marble" *Edelgrau* (see Chapter 2.9) for door panels, window sills and flooring

The Granite Super Bowl in front of the museum is a carved and polished erratic block designed and produced by Christian Gottlieb Cantian. It was one of the first machined stone objects and caused a sensation in Berlin. Because of its diameter of nearly 7 m, it was exposed in front of the museum in 1830. The origin of the granite was identified as *Karlshamn Granite* from South Sweden.

New Museum (Neues Museum)

Architect: Friedrich August Stüler (1800–1865)

Natural stones outside: Elbe Sandstones with the types *Postelwitz* and *Posta* (see Chapter 2.8) for the base, ledges, window walls, columns, pilasters, architrave and figurative building jewelry; partly repaired with Silesian Sandstones from Zbylutow (Late Cretaceous, Poland).

Figure 1.2.3 Main front of the New Museum to the Colonnade Courtyard.

Natural stones inside:

*   *Rüdersdorfer Muschelkalk* (light grey Middle Triassic (Lower Muschelkalk) limestone from Brandenburg, near Berlin) used for the basement, visible inside exhibition rooms of the basement
*   Carrara Marble (white Lower Jurassic marble from Toscana, Italy) with the varieties *Ordinario*, *Finochioso* and *Fosse de Zechino* for columns (shafts respectively bases and capitals)
*   *Campan Melange* (dark red Upper Devonian limestone from the Pyrenees, France) for columnar shafts; repaired with Saalburg "Marble" from the quarry Vogelsberg (see Chapter 2.9)
*   *Großkunzendorfer Marmor* (light grey Precambrian marble from Silesia, Poland) for flooring and original staircase (still remained on the last steps and landing at the 3rd floor)
*   Lahn "Marble" *Unika* (see Chapter 2.9) for floor tiles in the vestibule

Old National Gallery (Alte Nationalgalerie)

Architects: Stüler (design of the exterior), Johann Heinrich Strack (1805–1880, planning of the interior decoration)
Natural stones outside:

Figure 1.2.4 Old National Gallery.

- *Nebraer Sandstein* (see Chapter 2.2) for the entire façade with eight Corinthian columns and the double-run front staircase; repair of staircase with *Schweinsthaler Sandstein* (Lower Triassic sandstone from Rhineland-Palatinate), *Lausitzer Granit* for the steps
- *Striegauer* and *Strehlener Granit* (grey Carboniferous granites from Silesia, Poland) for the base of the building
- *Jaune Monton* (beige oolitic Middle Jurassic limestone from Dep. Yonne, France) for the relief frieze in the column porch

Natural stones inside:

- *Solnhofener Kalkstein* (see Chapter 2.11) for floor tiles in the vestibule on the passage to the Large Transverse Hall as remain of the original flooring pattern
- Saalburg "Marble" *Altrot* (see Chapter 2.9) for floor tiles in the western part of the Justi-Cabinets (small exhibition rooms) and in the Small Transverse Hall

Bode Museum

Architect: Ernst Eberhard von Ihne (1848–1917)

Figure 1.2.5 Bode Museum, entrance and west façade.

Natural stones outside:

* Silesian Sandstones (Late Cretaceous, Poland) with the types *Wünschelburger Sandstein* (reddish-brown patinated sandstone from Radkow, Gory Stolowe) for the curved entrance front in the northwest; *Friedersdorfer Sandstein* (light grey sandstone from Lezyce, Gory Stolowe) for half columns and pilasters; *Rackwitzer* and *Warthauer Sandstein* (Rakowice and Wartowice, Boleslawiec region) for façade fields, cornice, architrave and drum
* *Lausitzer Granit* (from Lusatia, Saxony) for the external staircase

Natural stones inside:

* Saalburg "Marble" *Königsrot* (see Chapter 2.9) for door panels
* *Bayerfelder Sandstein* (greenish grey Permian sandstone from Rhineland-Palatinate) for walls and columns in the Basilica
* *Untersberg* (beige, Late Cretaceous) for flooring and stairs
* *Grauschnöll* (grey, Lower Jurassic) for stringers (both limestones from Salzburg area, Austria)
* *Rosso Verona* (red Middle Jurassic nodular limestone from Veneto, Italy) for door panel and flooring
* Carrara Marbles *Ordinario* and *Bardiglio* and *Großkunzendorfer Marmor* for flooring

Figure 1.2.6 Pergamon Museum, main façade at *Kupfergraben*.

Source: 20130201_4273 PMU_Aussenansicht vom Kupfergraben, nach Nordost © Peter Thieme.

Pergamon Museum

Architects: Alfred Messel (1853–1909, design 1907), Ludwig Hoffmann (1852–1932, continuation after Messel's death with some alterations)
   Natural stones outside:

> Bavarian Muschelkalk, especially the varieties *Gaubüttelbrunner* and *Sellenberger Muschelkalk*, partly repaired with *Oberdorlaer Muschelkalk* from Thuringia (see Chapter 2.10)

Natural stones inside:

- Saalburg "Marbles" *Rot, Altrot* and *Edelgrau* for flooring
- Jurassic limestones (see Chapter 2.11), especially the varieties *Jura Gelb* and *Jura Gelb Gebändert* for flooring
- *Gauinger Travertin* (light brownish Tertiary limestone from Baden-Württemberg) for flooring and steps
- *Mühlhäuser Travertin* (light brownish Holocene limestone from Thuringia) for flooring

Figure 1.2.7 James Simon Gallery.   *Figure 1.2.8* Colonnades at the
                                              riverside.

- *Estremoz* (white-reddish Cambrian marble from Évora, Portugal) for the stairs

James Simon Gallery and colonnades

Architect of the James Simon Gallery: David Chipperfield
    Natural stones outside:
Lengefelder Dolomitmarmor (white Cambrian dolomite marble from the Erzgebirge Mts., Saxony) as surcharges in the cast stone for façades and colonnades (also used for complementary restoration of New Museum)

    Natural stones inside:

- *Crailsheimer Muschelkalk* (see Chapter 2.10) for flooring
- *Thassos* (white banded Mesozoic dolomite marble from Greece) for translucent natural stone-glass composite elements

The colonnades of the new entrance building take up the historical colonnades, which had been built in two periods: since 1853 with Elbe Sandstones (*Postelwitzer*) and since 1876 with Silesian Sandstones (*Rackwitzer* and *Warthauer*).

1.3 Carolingian Westwork and Civitas Corvey

Jochen Lepper

Preface

Situated in the interior core zone of the UNESCO Site, the Carolingian part of the Westwork and the archaeological monuments of the Carolingian Civitas [Dei], representing the City of God, still hidden in the underground of the Baroque monastery complex, are exclusively accredited as UNESCO World Heritage objects in 2014 (Ringbeck et al. 2012). All more recent buildings, including the Romanesque heightening of the Westwork and of the flanking towers with their even younger pinnacles, the Baroque hall church and all the other Baroque buildings of the monastery complex are not registered as UNESCO subjects.

History of the buildings

The central element of the medieval monastery complex was the Carolingian church, built in two phases between 822 and its final consecration 885. Changes in the 12th century affected the Carolingian Westwork by removing the squat middle tower and heightening both lateral towers, as well as inserting a transverse bell-story in-between. The pinnacles of both flanking towers were constructed around 1600 (Figure 1.3.1). After the demolition of the medieval church during the Thirty Years' War, a new Baroque hall church was erected (1667–1671), which in general follows the external outline of the Carolingian building (Lobbedey 2007). Later changes (in 1939 and 1965) made to the Westwork partly restored the original structure of the Carolingian interior.

Description

Westwork: The original Carolingian church building was characterized by a group of three square towers overtopping the Westwork, which was originally covered by plaster, supposed to be painted but now removed. The western façade, which appears as one massive, uniform building block, is characterized by a jutting central multi-story porch, three arches of the entrance hall and two stories of round-arched windows, whereas the flanking towers are opened by only a few slit windows. Romanesque

Figure 1.3.1 Façade of the Carolingian/Romanesque Westwork.

Source: Lepper and Ehling (2018a).

alterations include the removal of the mighty central tower and the heightening of the two lateral towers with their arcades, as well as the transverse bell-story, with their two rows of twin arcades (Figure 1.3.1).

In contrast to the massive, uniform building block of the western façade, the interior of the Carolingian Westwork incorporates a complex multi-story building composed of a central room on the ground floor, structured by stone facing columns and pillars and sidewise arcades (Figure 1.3.2). Situated above, the main room, known as the St. John's choir (Figure 1.3.2), is surrounded by two-story side chambers and stands out due to its original Carolingian wall paintings, plaster and fragments of colored architectural ornamentation (Claussen & Skriver 2007; Ringbeck et al. 2012). The preserved baroque hall church, built between 1667 and 1671, replaced the original Carolingian Basilica.

Civitas Corvey

Nearly the whole medieval monastery complex, except the church's Westwork, was demolished during the Thirty Years' War. Archaeological

Figures 1.3.2 Upper and ground floor of the Westwork.
Source: © Kulturkreis Höxter-Corvey GmbH, Peter Knaup.

findings and limited local excavations enabled the delineation of the Carolingian monastery complex, congruent to the extension of the Baroque complex. Details of the building stones recovered from the medieval stoneworks have not been reported (Ringbeck et al. 2012).

Natural building stones

As a whole, the currently bare stonework of the western façade is typified by at the most roughly trimmed, flat layered quarry stones of the thin-bedded to laminated *Roter Wesersandstein* (see Chapter 2.1). Reflecting this sedimentary thin bedding, the layered stonework, unique to the Westwork, displays a distinctive joint pattern of long stretching horizontal joints along the x-axis and relatively short perpendicular joints on the y-axis, framing each individual slab-shaped building block. This distinct fabric of the stonework, with its specific joint pattern, does not exist in any of the numerous medieval churches in the surrounding area. It perfectly reflects the local disposability of appropriate quarry stones nearby and was used in the same manner for the Romanesque heightening of the Westwork.

However, in the upper portions of the Westwork heightened in Romanesque times, a grey variety of the Weser Sandstone (see Chapter 2.1) was additionally used. This variety appears here exclusively as precisely masoned dimension blocks, as ashlars in the inter-arcade pillars, as exposed quoins and as fine-crafted columns with their capitals in the arcades. The same *Grauer Wesersandstein* can also be seen in the interior of the Westwork, though used only for precisely masoned rectangular dimension ashlars, as fine-crafted cylindrical shaped portions or monolithic columns with their elaborate capitals. Now, when the covering of former paint, limework or other coatings had been removed, they currently exhibit their original bare stone face. The restricted usage of the *Grauer Wesersandstein* in the medieval Westwork required some logistics for the transport from its outcropping area, some kilometers away in the Weser valley upstream of the river.

The homogenous massive variety of the *Roter Wesersandstein* was used externally and internally and is processed manually in the same manner as the *Grauer Wesersandstein*, but generally not used as raw quarry stone.

The younger saddle roof between the two flanking towers, as well as all the other roofs, is covered by laminated, fine-split *Roter Wesersandstein* (well-known as "*Sollingplatten*", Figure 1.3.3).

State of preservation of the building stones

All bare stone faces exhibited today were previously coated by plasters, limeworks or other paints that have now been widely removed. In general terms, all the now apparent sandstone surfaces appear not to have been damaged seriously by weathering. Disintegration of the otherwise

Figure 1.3.3 Assemblage of roofs covered by fine split so-called *Sollingplatten*.
Source: Lepper and Ehling (2018a).

solid sandstones can only be detected at some places where sulfuric acid was applied in former times to decoat the sandstone of any paints or plasters or outside where de-icing salt was spread in winter. Up to now, no systematic survey of the preservation of the building stones has been conducted and should be considered as a basis for any decisions regarding stone conservation measures.

Measurements for preservation

From 1947 up to now, several restoration phases have been executed at the Westwork, mainly to conserve the structure of the building, including the stabilization of the tilted towers, as well as to reveal and conserve the internal Carolingian and younger wall paintings, the historical plasters, Sinopias of stucco figures and fragments of architectural color ornamentation. Secondary damage caused by injections of lime-cement to stabilize the towers has induced additional cracking, followed by humidity impregnations and further loss of the original external plaster. Up to now, the cause of the inclination of the two towers of the Westwork has not yet been explained and urgently requires investigation. Up to now, no further measures have been declared necessary to protect the building and its ornamental stone substance, both inside and outside of the Westwork, from weathering and damaging deterioration (Ringbeck et al. 2012).

1.4   Cologne Cathedral (Kölner Dom)

Esther von Plehwe-Leisen and Hans Leisen

Introduction

Cologne Cathedral was built during a very long construction period from 1248 to 1880 with a break of ca. 320 years. Natural stone weathering and men-made decay always required measures for restoration and stone conservation. A total of 773 years of Cologne Cathedral construction and safeguarding history explain the extreme variety of building stones used. Besides, there are special natural stones used for stone sculptures and decorations. Different stones show dissimilar weathering behavior and need individual conservation concepts. This results in very complex questions for stone preservation. The work of the cathedral workshop is therefore a great challenge. In 1996, Cologne Cathedral was inscribed on the UNESCO World Cultural Heritage List. The monument is a magnet for visitors with ca. 6 million persons per year.

Figure 1.4.1 The western façade of Cologne Cathedral.

Figure 1.4.2 View of the flying buttresses on the northern nave made of Schlaitdorf Sandstone with varying weathering behavior of the stone blocks.

Construction history

The Cologne Cathedral of today is the result of about 770 years of building activity with a long interruption of ca. 300 years (Figure 1.4.1). A large Carolingian church complex stood on the site of the cathedral since about 800. However, when the Archbishop of Cologne, Rainald von Dassel, brought the holy relics of the Three Magi to Cologne in 1164, a new large church had to be built for the expected boom of pilgrims. Thus, the foundation stone was laid for the Gothic cathedral in 1248. In this first construction phase, the choir, the naves up to the height of the side aisle capitals, and the first two floors of the south tower up to the height of 59 m were built. The choir was consecrated in 1322 (Wolff 1974; Schumacher 2004).

Building work on the cathedral continued until the twenties of the 16th century, after which the building activity stopped. The reasons for this were lack of money due to a decline of trade with sacred relics and indulgences, but also due to recurring plague epidemics.

Almost 600 years after the laying of the foundation stone, construction was restarted in 1842. The completion of large cathedrals became a national task (Nipperdey 1968). During the French occupation of Cologne, 1794–1814, the cathedral was temporarily used as a magazine for the French army. Even prisoners of war were housed here (Wolff 1974). After the liberation of the cathedral, extensive safeguarding measures had to be taken at the medieval building (Schumacher 1993). Many stone parts had already suffered greatly during the centuries because of weathering and neglecting the building. The roofs had to be repaired, stones had to be replaced and complete building components had to be removed for safety reasons.

During the continuation of the construction, the cathedral was completed and decorated with sculptures. The rediscovered medieval façade plan (*Fassadenriss*) of the western façade was largely followed during construction.

The third phase happened between the First and Second World Wars. Weathered stones were exchanged or reworked by stonemasons. The destruction of the rock was particularly dramatic at the choir's buttress, where large parts of the medieval Drachenfels Trachyte had already been replaced by other rocks in the first half of the 19th century.

After WWII, the next phase of restoration began, which is still going on today. The war damage had to be repaired and large areas of stone were replaced again using additional kinds of stones.

Building stones of the Cologne Cathedral

For the individual construction and restoration phases, an extremely large number of different stones have been used at the Cologne Cathedral. Wolff (2004) mentioned a total of 50 different natural stones for building components and construction-bound decoration and sculptures. Besides, other stones were used for the valuable interior decoration with sculptures, tombs and pillar figures (Plehwe-Leisen et al. 2019). The interior decoration also changed during the Renaissance and baroque periods and thus, stones such as alabaster or black and colored limestones were introduced.

First phase of construction: 1248–ca. 1520

The stone selection during the first construction phase followed easy transport conditions. The quarries of the trachyte on the Drachenfels Mountain near Königswinter and the basalt columns from Unkel were situated around 44 km by boat up the Rhine directly above the river. Drachenfels Trachyte (see Chapter 2.14) was used as the general building stone for walls, arches, pillars, flying buttresses, window traceries, gargoyles, etc. Basalt columns were needed to build the foundations (Figure 1.4.3). The remaining stones from the Roman city of Cologne were even available directly on-site and could be reused. Only the transport of the light tuff for the vaults from the East Eifel volcanic field was a little more complicated.

For decoration and sculptures, the range of natural stones was larger. Besides reused stones – limestone from Norroy-lès-Pont-à-Mousson in Lorraine, Buntsandstein from Trier, marble from Carrara – Weibern Tuff from the Eifel (see Chapter 2.15), limestone from Wallonia and the Baumberge Sandstone from the Münsterland region (important sculptural stone, Figure 1.4.4) were used (Wolff 2001; Schumacher 2004; Plehwe-Leisen 2007a, 2019; Bergmann et al. 2019).

Second phase of restoration and completion: 1816–1880 (subsequent work until 1902)

The original building material, the Drachenfels Trachyte, was only available to a very limited extent in that time, and finally, the extraction stopped completely. Therefore, this phase was characterized by a permanent search for the perfect building stone for the completion of the cathedral (Schumacher 1993). Rocks from the surrounding area such as latites from Stenzelberg, Wolkenburg or Berkum, and basalts and tuffstones from the Eifel (Niedermendig, Mayen, Hannebach; Brohltal, Weibern) were exploited. At least 11 different sandstones from several parts of Germany were chosen. This is why this construction phase is often referred to "the sandstone age" of the Cologne Cathedral. Schlaitdorf Sandstone was used for the

Figure 1.4.3 Boreholes into the foundation of Cologne Cathedral reveal the construction technique and the materials used (B: basalt, T: tuffstone, M: mortar).

construction of large areas like the finalizing of the naves and flying buttresses (Figure 1.4.2). After completion of the railway line from Hannover to Cologne in 1848, the quartz sandstone from Obernkirchen (see Chapter 2.6) became the most important new building stone of the cathedral e.g., for the imposing towers. The other sandstones originate from Heilbronn, Flonheim, Staudernheim, Udelfangen and the Baumberge region. France, a country rich in limestone, supplied most of the stones (Hordain, Caen, Crazannes, Savonnières) for the architectural ornaments and sculptures (Figure 1.4.5) (Wolff 1972; Schumacher 1993).

Third phase of repair and exchange: early 20th century until WWII

The "limestone age" of the cathedral began. Large parts of the choir gallery, the pinnacles and flying buttresses were reworked in Krensheim Muschelkalk (see Chapter 2.10). Hidden areas, however, were renewed with the more resistant Eifel basalt according to doubts about the durability of the Muschelkalk limestone. Additionally, the French limestones from Savonnières-en-Perthois and Euville were used (Wolff 1972; Schumacher 2004).

Fourth phase of repair, exchange, and conservation: 1945 until today

After the WWII, serious war damages had to be repaired first. Following the good experiences with weathering-resistant basalt, the basalt from Londorf near Giessen was the preferred replacement material between 1952 and 1998 (Figure 1.4.5) (Wolff 1972). A limestone from Tercé in France was used for the replacement of degraded sculptures, canopies, and architectural decoration since 1989. Doubts about its durability led to the prophylactic use of acrylic resin full impregnation after the completion of sculptures and canopies. In exchange for the weathered Drachenfels Trachyte, trachyte from Monte Merlo (Italy) has been used since 2005 (DBU Project 2006), and sandstones from Radkow/Poland and Bozanov/Czech Republic are replacing the Schlaitdorf Sandstone since 2003 (Schumacher 2004).

Weathering

The weathering of rocks and the possibilities of their preservation have attracted the interest of many researchers since the late 19th century (compilation in Plehwe-Leisen 2007b; Graue et al. 2011).

Each of the stones used in Cologne Cathedral weathers in its own way requiring individual conservation concepts. Four, particularly important, building and sculptural stones are described in more detail next. All of them reacted sensitively to the air of Cologne, which was heavily polluted with sulphur dioxide until the last decades of the 20th century. Damaging salts developed, putting huge stress to the stone structure (Kraus & Jasmund 1981; Plehwe-Leisen et al. 2015).

The most important building stones of Cologne Cathedral are the Drachenfels Trachyte and the Schlaitdorf Sandstone. The original stone, the trachyte from the Drachenfels may show a heavy deterioration of the near-surface zones but in most cases, the stone interior below is still in good condition. As a consequence of stone replacements at the cathedral, the Drachenfels Trachyte is often surrounded by new "stone neighbors". It is particularly sensitive to these new influences and the degradation can accelerate. For example, the Krensheim Muschelkalk provides Ca^{2+} ions for the crystallization of harmful salts, and the sandstone of Obernkirchen shows different water absorption properties. It remains to be seen whether the use of the dense trachyte from Monte Merlo as a replacement stone will also lead to increased damaging of the original Drachenfels Trachyte. Schlaitdorf Sandstone used for Cologne Cathedral was obviously exploited in various quarries (Werner et al. 2013). While the building blocks with a siliceous-kaolinitic binder are resistant to weathering, the blocks with a carbonatic binder deteriorate particularly quickly (Figure 1.4.2). Nevertheless, the sandstone shows other deterioration patterns than the Drachenfels Trachyte. Some parts are so heavily destroyed that they have to be replaced by newly cut stone pieces.

The particularly valuable St. Peter's Portal with the medieval sculptural decoration is made of carbonatic sandstone from the Baumberge (Figure 1.4.4). Here, the sulphur dioxide-containing air of Cologne led to severe damages, too. The sculptures had to be replaced by copies between 1972 and 1980.

In the 19th century, more than 1,000 sculptures were produced for the completion of the Cologne Cathedral. The Savonnières limestone from Lorraine was preferred for this purpose (Figure 1.4.6). This limestone is less susceptible to weathering in clean air areas, as it forms a protective layer on the surface. In environmentally polluted Cologne, in the rain shadow, the sculptures are usually well preserved, but the exposed parts are often very heavily weathered (Plehwe-Leisen et al. 2015).

Figure 1.4.4 The precious stone carvings of the medieval Peter's Portal of Cologne Cathedral are made of Baumberge Sandstone.

Figure 1.4.5 The weather-resistant Londorf Basalt was used for stone exchange in the fourth phase.

Conservation and maintenance

The construction and preservation of Cologne Cathedral is in charge of the Cathedral Workshop (*Dombauhütte*) since 1842 with brief interruptions. Since 2018, the "Tradition of Cathedral Workshops" (*Bauhüttenwesen*) has been included in the list of Germany's intangible cultural heritage. In December 2020, the *Dombauhütte* and 17 other cathedral workshops from five European countries, have been inscribed in the international "UNESCO Register of Good Safeguarding Practices".

The Cologne Cathedral Workshop largely cultivates traditional craftsmanship. For stone preservation, it has mainly followed stonemasonry strategies like the reworking of the stone surface or stone exchange. Gradually, however, conservation approaches are now also coming into play. Since 2013, a small team of restorers has been supporting the stonemasons and architects of the Cathedral Workshop in matters of conservation. While in the case of stone replacement the selection of stone for new pieces is the most urgent task, in the case of stone conservation it is absolutely essential to have a precise knowledge of the properties, conditions, and weathering risks of the "stone patient" (Figure 1.4.7).

Since the 1970s, intensive research has been conducted to the influence of air pollutants on stone degradation at the cathedral. Various commercial stone conservation agents were tested and verified in exposure experiments (Luckat 1975; Mirwald et al. 1988; Plehwe-Leisen 2005; Plehwe-Leisen et al. 2007). Projects to investigate and develop

Figure 1.4.6 Nearly all sculpture decoration of the Cologne
Cathedral dates from
the 19th century like the
two kings of the western
façade. Around 600
sculptures and decora-
tion items were made of
Savonnières limestone.

Figure 1.4.7 Conservation work in
the sky executed by the
Cathedral Workshop.

conservation concepts for selected building stones of the Cologne Cathedral focused on deterioration factors and conservation materials and
strategies (DBU project 2006, EU Nanocathedral Project). The preservation of an enormous number of very different stone materials requires
many individual concepts (Plehwe-Leisen 2004). There is no off-the-peg
conservation that is successful for all stone materials and all weathering
patterns. Nevertheless, the new approach of exchange and transfer of
knowledge with other workshops focusing on stone conservation instead
of stone replacement is very promising.

1.5   *Bergpark* (Mountain Park) Wilhelmshöhe

Enno Steindlberger

The buildings in their historical context

The landscape garden of Wilhelmshöhe is located west of Kassel and is
considered Europe's largest artificial park of its kind. The monumental
fountains and the statue of Hercules give an impressive example of the
landscape architecture of European absolutism. It is a unique baroque
arrangement in which different movements in garden architecture, art

and technical history can be admired up to now. Therefore, it has been listed as a World Heritage Site in 2013. The Landgraves and Prince Electors of Hessen-Kassel started to design the park in 1696. In the following 150 years, it was continuously extended, so that the ensemble nowadays consists of a multitude of smaller and larger buildings. The main attractions of this site include the Hercules monument with cascades, fountains and grottos, the Lions Castle and Wilhelmshöhe Castle (Figures 1.5.1–3).

Hercules and cascades

High up, situated on the edge of the Habichtswald Mountains, the Hercules statue is enthroned as a landmark above the city of Kassel. The entire monument consists of an octagon as foundation with its courtyard, the pyramid and the crowning copper figure of Hercules. The period of construction was from 1704 to 1717 under the direction of the Landgrave Karl of Hessen-Kassel. He was able to acquire the famous Italian architect Giovanni Francesco Guerniero. The building has a height of 71 meters and represents the upper end of the cascade system, which was constructed down the slope, forming a visual axis (Figure 1.5.1). In the summer months, the "water games" attract numerous tourists. The water cascades downhill over multiple steps through a waterfall and an aqueduct, ends at the pond at Wilhelmshöhe Castle, where it rises as a large fountain.

Lions Castle (Löwenburg)

The artificial ruins of the *Löwenburg* were designed to imitate a romantic medieval knight's castle (Figure 1.5.2). This building complex consists of a tournament site, a castle garden and a vineyard. It was built between 1793 and 1801 under the aegis of the Landgrave Wilhelm IX of Hessen-Kassel. The later Elector Wilhelm I conceived it as a pleasure palace, and the interior comprises manorially furnished living quarters. The armory with weapons and knight's armor from the 16th and 17th centuries as well as the castle chapel are worth to note.

Wilhelmshöhe Palace (Schloss Wilhelmshöhe)

The palace was built in several stages from 1786 onward. Two former lateral wings (Figure 1.5.3) were enlarged by a central building and connected later on to form a building unit. Especially the middle parts of the

Figure 1.5.1 View from the Hercules down the cascade to the palace and Kassel-Wilhelmshöhe.

Figure 1.5.2 The Lions Castle looks like a medieval knight's castle.

palace were completely destroyed during World War II. A reconstruction with a reutilization as a museum took place between 1968 and 1974.

Natural building stones

The building material used in the park is mainly tuff (see Chapter 2.16), which is well suitable for the purpose as grotto stone. During the

Figure 1.5.3 Wilhelmshöhe Castle, one of the lateral (northern) building wings made of tuff.

construction phase, the required quantities of tuff came from the immediate vicinity of the construction sites (Steindlberger 2003, 2011). All historical quarries are dilapidated and exploited nowadays. Only in the 1980s and from 2007 onward, material of a special excavated quarry located in the nearby Drusel Valley was available for restoration purposes.

In addition to the mainly used tuff, some reddish to yellowish-white sandstones (Buntsandstein) had been used, for example, at the central building of *Schloss Wilhelmshöhe* or within some sections of the Lions Castle. The sandstones originated from a quarry area west of Kassel. Nowadays, comparable, but more violet or greyish-white sandstones (Weser Sandstone, see Chapter 2.1) are available from a quarry area near Bad Karlshafen. The use of these sandstones guarantees a more durable and high-quality ashlar masonry. Furthermore, basalt lava was used to replace damaged tuffstones, for example for the slabs or steps on the platform of the octagon. Initially, a basalt lava from the Vogelsberg Mountain (Londorf type) had been placed; nowadays, due to its better availability, a similar basalt lava from the Eifel region serves as replacement material.

State of preservation

The Hercules building showed already serious structural damages since its construction. They are caused by static-constructive misjudgements regarding the soft and susceptible tuff. Due to the inconsistent and subsiding geological subsoil directly on the sloping edge of the terrain, structural improvements had to be carried out subsequently. The most recent measures included fundamental strengthening concepts (Haberland & Huster 2007; Sander 1981).

Compared with all the used building materials in the park, tuff is a low-strength rock and, accordingly, a building material that is very susceptible to weathering. Especially, building parts exposed to high moisture loads are often affected. The weathering processes are associated with an unfavorable pore distribution, resulting in a high water absorption ability. Due to freeze-thaw processes, the rock fabric disintegrates, leading to cracking and crumbling. Under the influence of alternating humidity, swelling and shrinking processes of clay minerals lead to tensions within the affected zones, first, and finally to characteristic damage patterns like scaling, flaking and sanding (Figure 1.5.4).

Conservation measures

For the reconstruction of masonry of large sections within the three buildings, tuff residuals from the recently operated quarry have been used up to now. Due to the lack of suitable stone material, additionally a cementitious artificial tuff has been developed. Designed as a replacement material, it was adapted to the original rock properties (Figure 1.5.5).

To preserve this unique cultural heritage and its historical masonry, conservation and restoration measures have to go on. However, this type of tuff has to be considered as a "problematic stone" (Steindlberger 2020a), and standardized conservation strategies are inadequate. As a result of several years of research, optimized means and methods can be presented by now (Steindlberger 2020b). In principle, formulated soaking agents, adapted to the specific characteristics of the tuff, are quite effective. They are based on a silicic acid ester (SAE).

Figure 1.5.4 Loss of the historical stone surface at the Lions Castle due to scaling.

Figure 1.5.5 Arrangement of natural (middle) and artificial tuffs as replacements, not finally processed.

1.6   Classical Weimar and Bauhaus in Weimar

Lutz Katzschmann

In the late 18th and early 19th centuries, the small Thuringian residence Weimar developed into one of the leading cultural centers of its time. The classical period of Weimar began with the work of Duchess Anna Amalia, the commitment of Christoph Martin Wieland and Gottfried Herder, the promotion of art by Duke Carl August, and culminated in the creative poet friendship between Friedrich Schiller and Johann Wolfgang von Goethe. Cosmopolitanism and humanistic striving characterize the literary works created during this period. The buildings and parks of that time represent the extraordinary cultural significance of Classical Weimar.

Since 1998, the World Heritage Site of the Classical Weimar includes 12 separate buildings and ensembles:

* Goethe's Residence with National Museum
* Goethe's Garden House
* Schiller's Residence
* *Wittumspalais* (Dowager's Palace)
* *Herzogin* (Duchess) *Anna Amalia Bibliothek* (Library) with the famous Rococo hall at its heart
* St. Peter and Paul (Herder church) with Herder House and Garden as well as the neighboring Old Grammar School
* Historical Cemetery with Ducal Vault
* Residence (City) Castle and Ensemble Bastille
* Park by the River Ilm with the Roman House
* Belvedere Castle and Park with Orangery (Figure 1.6.1)
* Tiefurt Mansion and Park
* Ettersburg Castle and Park

The building materials used were mainly rocks from the immediate vicinity of the city. These include the Quaternary travertines of Weimar and Ehringsdorf, the limestones of the Upper Muschelkalk from the surroundings of the city and the greyish-yellow and reddish-brown sandstones of the Middle Buntsandstein, quarried near Bad Berka and Tonndorf south of Weimar.

The rising masonry of the City Castle and the other buildings mostly consist of limestone of the Upper Muschelkalk, brick, and more rarely travertine and sandstone of the Lower Keuper. Stone-facing masonry is relatively rare. Examples are the lower part of the castle tower (limestone),

Figure 1.6.1 Belvedere Castle.

the Templar's House in the park on the river Ilm (travertine) and the south wing of the City Castle built in 1911/13 with White Main Sandstone (quarried near Eltmann/Bavaria).

The weathering-resistant Weimar Travertine was often used for building bases. Stone masonry components that are more complex to process are mostly made of the sandstones of Bad Berka and Tonndorf south of Weimar. Typical examples are portals, jambs and cornices at the City Palace and Goethe's House (Figure 1.6.2), columns and plinth masonry of the Roman House or columns, window jambs and staircases of Belvedere Palace (Figure 1.6.1). Slate roofs consist of dark blue-grey Thuringian roof slates (among others Herder church, castle tower). The castle building records show the use of Lehesten Roof Slate since 1780 (Seidel & Steiner 1988). Local stones as well as materials from more distant locations were used for sculptures, monuments, elements of the park architecture and gravestones.

Examples are the lions in front of the west portal of the City Palace that are made of Thuringian Rhaetian Sandstone (Figure 1.6.4, see Chapter 2.5) and the so-called Schlangenstein (Snake stone) in the Park on the river Ilm, which is made of Tonndorf Buntsandstein. The monuments to Franz List and William Shakespeare (Figure 1.6.3), erected around 1900, are made of Carrara Marble and stand on pedestals made of Jurassic limestone and blue-grey Kösseine Granite (Fichtelgebirge/Bavaria), respectively. The most famous Weimar monument, the Goethe-Schiller monument, rests on a pedestal of Raumünzach Granite (Bavaria). Travertine, sandstones from Buntsandstein as well as Rhaetian Sandstone

Figure 1.6.2 Goethe's Residence (portal and window jambs –
red Tonndorf Sandstone).

Figure 1.6.3 Shakespeare monument
(Carrara Marble, pedestal
Kösseine Granite).

Figure 1.6.4 Lion sculpture in front of the west portal of the City Palace
(Thuringian Rhaetian Sandstone).

(see Figure 2.5.3) and marble were used for the gravestones and tombstones in the historic cemetery.

Bauhaus sites in Weimar

The Bauhaus School, existing between 1919 and 1933, revolutionized aes-
thetic and architectural concepts and practices. It represented a completely
new fusion of art and craftsmanship. The Bauhaus is the most important
institution of art and architecture in the 20th century. UNESCO recognized

the experimental building "*Haus am Horn*" along with the two art school buildings in Weimar and other ones in Dessau as a World Heritage Site in 1996.

The main building of the Bauhaus University Weimar was designed by Henry van de Velde. Today it houses the President's Office as well as the Faculty of Architecture and Urbanism. It was built in two construction phases for the former Grand Ducal Saxon Art School between 1904 and 1911 and later integrated in the Bauhaus School founded by Walter Gropius in 1919. The Art Nouveau building with plastered brick walls rests on a base of Ehringsdorf Travertine. Window sills and lower parts of window jambs were made of Cretaceous Elbe Sandstone (see Chapter 2.8) and Thuringian Rhaetian Sandstone (see Chapter 2.5). Particularly impressive are the bar-free curved windows of the top floor and the elliptical main staircase in the interior.

The Van-de-Velde-Building of the Bauhaus University Weimar was designed by Henry van de Velde for the Grand Ducal Saxon School of Arts and Crafts. It was constructed from 1905 to 1906 and used by the Bauhaus School of Weimar between 1919 and 1925. Now it houses the Faculty of Art and Design. The angular construction corresponds in some features with the façade of the opposite main building. Here, too, the base is made of travertine. The south gable of the building with an arch made of travertine is remarkable. Additionally, visible steel girders rest on travertine ashlars (Figure 1.6.5). The interior design amazes visitors with its

Figure 1.6.5 South gable of the Van-de-Velde-Building.

unconventional lighting of the stairwell and three restored wall paintings, originally made by the Bauhaus master and artist Oskar Schlemmer.

1.7 Naumburg Cathedral (*Naumburger Dom*)

Robert Sobott

Introduction

Naumburg on the river Saale is located about 14 km from the center of the Saale-Unstrut-Triasland Geopark in the State of Saxony-Anhalt in Germany. For centuries, sandstone and limestone were quarried in this region and used for the construction of sacral and profane buildings. The sandstones belong to the Bernburg, Hardegsen and Solling Formation of the Lower and Upper Buntsandstein, respectively. The workable limestone layers of the Schaumkalk Formation of the Lower Muschelkalk, which are outcropping on the slopes of the Unstrut River valley between Freyburg and Balgstaedt, furnished the building material for Naumburg Cathedral. The extraction of *Schaumkalk* from a quarry near Balgstaedt is testified by a document from the year 1152 issued by the monastery Pforta in which Ulrich of Balgstaedt granted Bishop Wichmann the extraction of stone from his quarries for the construction of the cathedral (*Balgstedter Chronik*). Koch et al. (1999) gave a detailed account of the petrographical and petrophysical properties of the so-called *Freyburger Schaumkalk* (see Chapter 2.10), the principal building stone of Naumburg Cathedral (Sobott 2013).

Building history and description

Naumburg Cathedral is one of the most famous medieval church buildings in Germany (Schubert 1997). It is situated about 25 m above the water level of the river Saale on a plateau of the geological formation Upper Buntsandstein. The buildings of the cathedral complex are arranged around the cloister (Figure 1.7.1) and comprise, starting in the north, in clockwise direction the cathedral, the Chapel of the Three Holy Kings, the Church of St. Mary and the south and west close. The entrance to the complex is through a gate building between the Chapel of the Three Holy Kings and the Church of St. Mary.

The late Romanesque and early Gothic cathedral with numerous later additions and reconstructions was built from about 1210 to 1260 and includes parts of the antedating early Romanesque building as was revealed by archaeological excavations in the years 1961–1965 (Leopold & Schubert 1972). The oldest remaining structure is the Romanesque part of the Eastern crypt that was integrated in the early Romanesque cathedral about 1160/1170.

The west choir with the rood screen and the famous portrait statues of the 12 cathedral founders are of utmost art historical significance (Figures 1.7.2 and 1.7.3). It was built about 1250–1260 by the so-called Naumburg Master, a stone sculptor and architect trained in France to whom also works in Noyon, Amiens, Reims, Mainz and Meissen are attributed. The statues of the cathedral founders are early Gothic colored sculptures carved in *Freyburger Schaumkalk*. They are exceptional because the sculptor did not fashion them in the stereotype manner characteristic of sculptures of the medieval chivalric-courtly culture but gave them individual looks expressed by physiognomy, gesture, attire and attributes (Figures 1.7.2 and 1.7.3).

The west rood screen that was likewise created by the workshop of the Naumburg Master separates the west choir from the nave. Eight reliefs of the rood screen expressively show the passion of Christ, beginning with the Last Supper in the South and ending with Christ on the cross in the North. The passion narrative is continued and highlighted with Christ on the cross, flanked by Mary and John in the center of the passage through the portal

Figure 1.7.1 East towers and inner courtyard.

Source: © Vereinigte Domstifter zu Merseburg und Naumburg und des Kollegiatstifts Zeitz, Bildarchiv Naumburg, photo: Matthias Rutkowski.

Figure 1.7.2 Statue of founder Countess   1.7.3 West choir (general view).
Berchta, west choir (*Frey-*
burger Schaumkalk).

Source: © Vereinigte Domstifter zu Merseburg und Naumburg und des Kollegiatstifts Zeitz, Bildarchiv Naumburg, photo: Matthias Rutkowski.

projection in the middle of the rood screen. The leaf frieze and the column capitals in the two blind niches right and left of the portal projection are carved as delicately from nature that the plant leaves and blossoms are easily recognizable as grapevine, ranunculus, field maple, mugwort, corydalis cava and hazelnut without difficulty (Figure 1.7.4).

The figurative works of the Naumburg Master include also the tomb of Bishop Dietrich II, the builder of the west choir, and a lectern shaped as deacon in the east choir. Further outstanding pieces of decoration are the stained-glass windows in the west choir, a stone statue of Saint Elizabeth in the chapel of the same name on the ground floor of the northwest

Figure 1.7.4 Leaf frieze, west choir, limestone *Freyburger Schaumkalk.*

Source: © Vereinigte Domstifter zu Merseburg und Naumburg und des Kollegiatstifts
Zeitz, Bildarchiv Naumburg, photo: Matthias Rutkowski.

tower, the wooden stalls in the east choir and numerous bronze and stone
epitaphs in the aisles.

Very particular is the use of *Wechselburger Garbenschiefer* (fascicular schist) as decorative panels in the west choir (Bachmann & Glässer
2015). On both sides of the choir are stalls below the statues of the
founders offering space for 16 clerics on each side. Rows of Romanesque semi-columns crowned by capitals with vine leaf decoration and
canopies made of *Freyburger Schaumkalk* (limestone), both dated to the
13th century, serve as dorsal. The dorsal panels between the columns
were made of fascicular schist. In a blaze in the year 1532, the stalls
caught fire wherefore a part of the canopies and the fascicular schist
panels had to be replaced in the 1870s. Because of the exposure to fire,
the original schist panels have now a reddish color and a rough surface
(Schubert 1997). The *Wechselburger Garbenschiefer* is a metamorphic
rock originating from the schist cover of the Saxonian granulite complex
and was quarried until the 19th century (Bachmann & Glässer 2015). It
has a silvery muscovite-rich groundmass and irregularly oriented 5–6 cm

long and up to 1 cm thick dark porphyroblasts on the schistosity planes. The porphyroblasts are most probably pseudomorphic replacements of staurolite by dark phlogopite mica.

Outstanding pieces of art and superb craftsmanship are not restricted to the interior of Naumburg Cathedral. The east façade of the Chapel of the Holy Three Kings displays an outstanding high medieval Epiphany sculpture group that shows Mary and the child and the Three Holy Kings. It was carved from *Freyburger Schaumkalk* (Figure 1.7.5).

The two eastern cathedral towers show in the upper octagonal section the transition from late Romanesque architecture with double-arched windows (biforium) to late Gothic architecture with blind tracery and baroque hoods with two-story lanterns on top. At the north façade of the nave, wall arches and corbels for vault ribs are the remnants of a demolished cloister. The gargoyles on the west choir also derive from the workshop of the Naumburg Master (Figure 1.7.6).

Of the towers on the west side with an architecture very similar to that of the towers of the cathedrals in Laon and Bamberg, only the northwestern tower was completed in the 14th century with three medieval floors above the late Romanesque square substructure. The piers in the lofty construction are not exclusively made from *Freyburger Schaumkalk* but some were also made from red sandstone (Buntsandstein). The completion of the southwestern tower took place in the second half of the 19th century (1884).

Figure 1.7.5 Sculpture group: the Three Holy Kings. East façade of the chapel of the same name.

Figure 1.7.6 Gargoyle, west choir, limestone *Freyburger Schaumkalk.*

Source: © Vereinigte Domstifter zu Merseburg und Naumburg und des Kollegiatstifts Zeitz, Bildarchiv Naumburg, photo: Matthias Rutkowski.

Weathering and restoration

The highly porous "*Schaumkalk*" limestone is prone to chemical weathering. Acidic rainwater containing dissolved sulphur dioxide reacts with calcite to form gypsum that is more soluble in water than calcite. A thin gypsum layer on the limestone is formed by this reaction, which eventually grows to a gypsum crust several millimeters thick. Dirt particles trapped in the gypsum layer give it a dark or even black color. Once the limestone surface has been transformed to gypsum, two things may happen: (I) from exposed limestone surfaces, the gypsum layer will be washed off by rain, or (II) in sheltered areas, the gypsum layer may grow and eventually be detached from the limestone surface (Neumann 1994). In both cases, the limestone suffers a material loss (Fitzner & Heinrichs 1992). The extent of damage may be assessed by nondestructive methods, for example ultrasonic measurements, and determines the appropriate restoration/conservation procedure. In most cases, the first step will be a cleaning of the limestone surface, that is, the removal of the gypsum layer. The full extent of damage will be visible and appropriate conservation measures can be scheduled after the removal of the gypsum crust. Controlled dry/wet abrasive methods are applicable for the cleaning of ashlars, but in the case of sculptures or inscribed stone tablets, a subtler method such as laser cleaning should be considered. This method can be applied with exceptional success to light-colored stone surfaces because laser light is highly absorbed and poorly reflected by the dark layer and the irradiated photon energy is largely converted to heat, which literally evaporates the gypsum crust.

In the last 30 years, a number of restoration projects were carried out. In the late 1970s, the northwest tower underwent major restoration work. The limestone and sandstone piers of the arcades on the three Gothic floors were completely overhauled. Each section of the piers was analyzed by ultrasonic measurements (Sobott 2009) and worked stones with visible damage or low ultrasonic velocities were replaced by new *Schaumkalk* material. As the *Freyburger Schaumkalk* is no longer quarried, limestone material with an outer appearance and petrophysical properties that come close to that of the original material must be used for the replacement of larger worked stones. In 2011, the Epiphany sculpture group was restored and a comprehensive account of the restoration work was given by Bauer-Bornemann (2012).

The inscription of Naumburg Cathedral to the UNESCO World Heritage List in 2018 set further restoration work in motion. In 2020, the cleaning of east choir facade started, which is accompanied by repointing measures and stone repair.

1.8 Trier – Roman and medieval World Heritage Sites

Friedrich Häfner and H. Wolfgang Wagner

Introduction

Nowhere else north of the Alps can you experience the Roman Age as authentically as you can in Trier. Here you will find the center of antiquity in Germany. Trier, the oldest German city, was founded as Augusta Treverorum and once was one of the largest metropolises of the Roman Empire.

The cityscape of Trier today is still characterized by monumental buildings dating from ancient times, with some of them among the best preserved of their kind. In 1986, nine Roman buildings and – as far as existent – their medieval supplements like the Cathedral, the Church of Our Lady, the Porta Nigra, the Imperial Baths, the Amphitheatre, the

Roman Bridge, the Barbara Baths, the Constantine Basilica and the Igel Column ("*Igeler Säule*", near Trier) were included in the list of UNESCO World Heritage Sites. One year later, the thermal baths on the cattle market were discovered during excavation works. Some of these monuments are described in more detail next (Zentrum der Antike 2020).

Porta Nigra

The Porta Nigra, the emblem of Trier, is the best-preserved Roman city gate north of the Alps. The gate was built in around 170 CE using approximately 7,200 blocks of stone and has been preserved until today, thanks to its solid construction. With the end of the Roman Empire and changing times, the city gate was used for several purposes. In the 11th century, it served as an abode for the monk Simeon, who lived as a hermit. Later on, the gate was converted into a church, which was another reason for its survival. In 1803 under Napoleon's rule, the church was dissolved and orders were given to restore its ancient design (Zentrum der Antike 2020). After Schumacher and Müller (2011), the Porta Nigra was built by using Kylltal Sandstones (see Chapter 2.3). This is proved by mason's marks, which indicate that the monumental rocks came from the open pit "*Pützlöcher*" in the Kylltal (valley) near the village Kordel. This open pit is situated about 12 km away from the Porta Nigra. At this place, the Romans first established a copper mine in the 2nd century CE, and subsequently, the mine was converted into an open pit to produce building stones (Figure 2.3.2) (Cüppers 1990).

The walls of Porta Nigra consist of huge blocks up to a length of more than 2 m, a height of about 0.6 m and a depth until 0.9 m. The blocks were arranged without using mortar and were fixed with iron cramps and backfilled with lead (Figure 1.8.2) (Löhr 2015).

Figure 1.8.1 Roman city gate "Porta Nigra" in Trier.

Figure 1.8.2 Huge blocks of light-colored Kylltal Sandstone (from Kordel) were used to build the Porta Nigra.

The façade shows dark to black colors, which were the reason to call the city gate Porta Nigra since medieval times. Auras (2014) described brownish and grey discoloration of the rocks' surface by increasing Fe-oxides and Fe-hydroxides as well as deposits of dirt, crusts of gypsum and microbiological growth of moss, green algae and lichen. Löhr (2015) is reporting on the influence of scorch marks. In 2013, a research project was established to investigate the alterations of the stone surfaces and to prepare restoration works.

Thermal baths on the cattle market (Viehmarkt-Thermen)

At this place, you can explore archaeological relics dating from the foundation of Trier to the end of World War II. The relics include the foundations of Roman houses, followed in the 2nd century CE by a major building, which was converted into public baths in the 4th century. Streets of houses and sewers, a medieval refuse pit and cellar rooms of the baroque Capuchin monastery were excavated from 1987 to 1994. The archaeological remains are surrounded by a protective glass shelter today (Zentrum der Antike 2020).

Völker (1999) especially examined the dimension stones of the *Viehmark-Thermen* and the amphitheater. He described five variations of carbonate rocks. The rocks vary in color (greenish-grey, grey, beige), porosity, content of glauconite, relics of crinoides, trochites, ooides and brachiopodes. The probable origin of the dimension stones are some Roman quarries in the Trier region (Völker 1999) (Figure 1.8.3). Stratigraphically, the rocks belong to the *Trochitendolomit, Linguladolomit* and the *Ceratitenschichten –* all Middle Trias (German Muschelkalk).

Figure 1.8.3 Interior view of the Thermal Baths on the Cattle Market in Trier.
Photo: M. Auras.

Cathedral (Dom)

With a history of worship and construction going back 1,700 years, the Cathedral of Trier is the oldest church in Germany. In its entirety, the mighty structure of the Cathedral is a compendium of European architectural and art history. According to historical sources, the mother of Emperor Constantine, Helena, is said to have given her house to the Bishop of Trier, Agritius, to be used as a church building, and brought the robe of Christ, known as the Holy Robe, back to Trier from a pilgrimage. Archaeological excavations beneath the Cathedral have indeed proved the existence of an ancient dwelling, while the safekeeping of the relic of Christ was documented for the first time in 1196. Renovation work at the first basilica was followed in the 4th century by the construction of a monumental church, which – with its four basilicas – was one of the largest church buildings of the 4th century. A large square section of the building was constructed around 340 CE and still forms the structural core of the Cathedral today. In the following years, the Cathedral was exposed to the vicissitudes of history. In the Middle Ages, there were numerous extensions and renovations, which are still largely preserved in the interior infrastructure of the Cathedral (Zentrum der Antike 2020).

The antique construction phases can be well recognized on the north wall via the different natural stones used in construction (Figure 1.8.4; Wagner et al. 2012): In the first square building around 350 CE, only dolomite stones (*Trochitendolomit*; Wagner 1991, Figure 1.8.4: A) alternating with Roman bricks were used. After an interruption in construction, about 380 CE to 390 CE the construction continued under the emperors Gratian and Valentinian II, which is still preserved today with brickwork made of red sandstone (*Pallien-Schichten*, Lower Buntsandstein, Triassic)

Figure 1.8.4 The north wall of the Cathedral shows the Roman and later construction phases based on different use of natural stones. (the figure is a montage of two different photos).

alternating with bricks up to a height of 25 m (Weber 2003; Jelschewski 2020; Figure 1.8.4: B, Figure 2.3.1). This building phase also includes the original 12 m high round columns of the interior, the relics of which lie in front of the southwest portal as the "cathedral stone" (*Domstein*). This quartz diorite originates from the "*Felsenmeer*" near Reichenbach in the Odenwald. Original fragments are exhibited in the "*Museum am Dom*", together with a reconstruction of part of a column, including the capital and abutment made from Auerbach Marble (marble, ?Devonian), also from the Odenwald.

The marble remains of the interior that were excavated in the 19th century also date from this period (Groß-Morgen 2020). They had been installed as floor patterns on the ground floor of the "Baden Chapel" in the north wing of the Cathedral cloister. Among the marbles are green diabase (Ruwertal, metabasalt, Devonian) and red, dark to light limestones with similarities to the incrustations used in the Constantine Basilica (Ruppiene 2021).

The Romanesque West-Work with choir predominantly built from dolomite building stones (partly reused by ancient predecessors) dates from 1040 to 1075 CE (dendrochronological dating by Zink 1980) (Figure 1.8.4: C). In addition, larger stones (red and light sandstones and yellow limestones) were reused as spoils from the previous Roman buildings, particularly in the portals and in the column galleries. These are primarily Kylltal Sandstone (see Chapter 2.3) and Jurassic Jaumont Limestone from Metz area (France) (Nancy-Metz 2020). In 1514–1515, the right tower (Greiffenklau tower, Figure 1.8.6) was raised by an additional story using light Kylltal Sandstone. Following the antique square building, the Late Romanesque east choir was added 1160–1196 CE with light and red Kylltal Sandstone (Figure 1.8.4: D). In the 14th century, the two Gothic east towers (Figure 1.8.4: E) were supplemented with red Kylltal Sandstone. Around 1700 CE, the baroque round "*Heiltum-skammer*" (Holy Robe repository) with mainly Udelfangen Sandstone (yellowish) (Lower Muschelkalk, Triassic) was added (Figure 1.8.4: F). All roofs of the Cathedral have been covered with roof slate since 1970 (*Moselschiefer*, Lower Devonian, see chapter 2.12) (Figure 1.8.4: G).

The rich interior includes works made of Kylltal Sandstone such as the archivolt with tympanum from 1180 CE at the portal to the *Lieb-frauenkirche* or the Renaissance pulpit (approx. 1570) (Figure 1.8.5). Most baroque altars and tombs, including the east choir, are made of so-called "*Buntmarmor*" ("Lahn and Eifel Marble" (see Chapter 2.9); polishable limestone, mostly coral and reef limestone, Middle Devonian). The dark columns of the large galleries and choir barriers were made of polished slate until around 1900 and were then replaced by black *Petit Granit* from Belgium (dark limestone, Carboniferous) (Jelschewski 2020). The new main altar from 1974 CE was made from a grey-green *Peperino* (trachyte tuff from Viterbo/Italy, Quaternary) with ornaments from Preonyx (alabaster/gypsum from Volterra/Italy, Tertiary) (Paulinus 2020).

The entire Cathedral was plastered at first. Since the 19th century, it has not been plastered outside, since 1974 also inside, so that the natural stone masonry can be seen. During the last extensive restoration 1960–1974, all natural stones were cleaned and replaced, if necessary. No major damage is known to date. In 2001, the Gothic Greiffenklau tower (Figure 1.8.6)

Figure 1.8.5 Renaissance pulpit of the Cathedral (approx. 1570 CE).

Figure 1.8.6 Gothic central building of the *Liebfrauenkirche* seen from the cloister. The Cathedral's Greiffenklau tower is in the right background.

was renovated with the reconstruction and manufacture of two new gargoyles (Bungert & Wirtz 2020). At the same time, the slate roof was newly covered. It had previously been covered with the Feller/Thommer Slate typical for Trier (Lower Devonian), which is no longer available today.

Our Lady's Church (Liebfrauenkirche)

The story of the *Liebfrauenkirche* began with Emperor Constantine the Great and his mother Empress Helena. From 316 CE, an early Christian church center with four basilicas (SW, SE, NW, NE) was built. The *Liebfrauenkirche* later emerged from the southeast basilica (approx. 335 CE), the later Cathedral from the northeast basilica (approx. 340 CE) via the square building (approx. 350 CE) (Weber 2017). A new central building of a church was built at the south side of the Cathedral with the outline of a 12-leaf rose (Rosa Mystica) in the finest French, highly Gothic, filigree architecture by builders coming from the Ile de France and the Champagne (Reims) between 1227 and 1269. Like its predecessor, it was consecrated to the mother of God (Zentrum der Antike 2020). At the same time and in the same style, the common cloister of the Cathedral and *Liebfrauenkirche* was built (Figure 1.8.6). In the new Gothic building – which was originally not stone-sighted – light, subordinate red Kylltal Sandstone and partly light yellowish Udelfangen Sandstone were used. The West Portal contains a few original sculptures, casts of

Figure 1.8.7 West Portal of the *Liebfrauenkirche.*

originals and modern new creations from Kylltal Sandstone, Udelfangen Sandstone and Jaumont Limestone (Jurassic; Figure 1.8.7).

The most magnificent portal in the interior is the Gothic North Portal, that is, the counterpart to the portal with the tympanum from 1180 CE in the cathedral. The archivolt in the *Liebfrauenkirche* is made of light Kylltal Sandstone (from Temmels/Saargau), with the sculptural parts (e.g., Typanum) being made from Jaumont Oolite (Jurassic) (Ehlen 2011). The *Liebfrauenkirche* also has many baroque altars and tombs made of light Kylltal Sandstone (from Kordel) and "Lahn and Eifel Marble" (see chapter 2.9). Extensive restorations were carried out in the 19th century, which gave the inside and outside of the church most of its current, stone-sighted appearance (Liebfrauenkirche 2020). The damage caused by World War II was only partially repaired. From 1985 to 2005, there was an external restoration with a replacement of the slate roofs (*Moselschiefer*, see chapter 2.12, Ehlen 2011). An exemplary interior restoration with extensive cleaning, replacement and sculptural addition of natural stones took place from 2007 to 2011 (Wagner & Wechsler 2009; Wagner 2012). The existing color versions and wall and ceiling paintings were cleaned and if necessary renewed. The church interior was also given a new floor made of Red Main Sandstone (*Roter Mainsandstein*, Lower Buntsandstein, Triassic). The north portal to the cathedral has just been restored. The restoration of the West Portal is still pending (Figure 1.8.7).

Roman Bridge (Römerbrücke)

The *Römerbrücke* (Figure 1.8.8), situated at the intersection of various roads, played an important role for the founding of the city of Trier and is still in use today. Timbers discovered in the bed of the river Moselle have undergone dendrochronological analysis and have been dated to 17 BCE. The first wooden bridge remained in existence for around 90 years and then was replaced by stone pillars and wooden arches. The present day *Römerbrücke* was reconstructed around 154–157 CE (Ostermann 2001), probably because the traffic routes were too narrow. In 1343, the wooden arches were replaced by sandstone arches (Cüppers 1990). In the 19th and 20th centuries, the traffic lanes have been widened due to the growing traffic volume. (Zentrum der Antike 2020). Seven of originally nine pillars remained and are still in use.

Mainly three different rocks were used to build the *Römerbrücke*. First, Rhenish Basalt Lava was transported from the quarry "*Hohe Buche*" near the village of Namedy, which is situated directly at the Rhine river (Wagner 2012; Schumacher & Müller 2011; Schaaff 2000). At this spot, the lava stream reached to the banks of the river and this place was ideally appropriated to extract the rocks and transport them by ship to Trier. The rock can be classified as a phlogopite leucite basanite.

Second, Blue Stone (*Blaustein*) was used. *Blaustein* is a collective term for blueish grey to blueish black limestones of the Rheinisches Schiefergebirge originated in the Palaeozoic Devonian and Carboniferous. *Blaustein* was extracted at least since Roman times. At the beginning of the

Figure 1.8.8 Roman Bridge in Trier, which is still in use; pillars built with Rhenish Basalt Lava, Blue Stone and sandstone in the 2nd century CE.

18th century, red sandstones (Buntsandstein, Triassic) were brought from the village Biewer near Trier to rebuild the arches of the bridge (Fachbach 2011).

1.9 Maulbronn monastery complex (*Kloster Maulbronn*)

Angela Ehling and Wolfgang Werner

Building history and specification

The former Cistercian monastery of Maulbronn is one of Europe's most complete and best-preserved medieval monastery complexes. It was founded in 1147 on an area of 145 hectares. It combines a multitude of architectural styles, from Romanesque to late Gothic, in one place – creating a unique atmosphere. Buildings in Renaissance and Historism

Figure 1.9.1 Tower and estate buildings of the monastery seen from the large courtyard.

Figure 1.9.2 Church with entrance hall, the so-called paradise, and fountain on the large courtyard. All buildings are made of Maulbronn Sandstone quarried nearby.

styles occur as well. Exceptional is that the whole medieval monastic infrastructure is preserved consisting of the monastery itself, farm buildings, wine fields, quarries and numerous fish ponds and water reservoirs for the mills including the necessary water channels.

In 1993, the monastery was inscribed on the list of UNESCO World Heritage Sites. With the following justification: "The Maulbronn complex is the most complete survival of a Cistercian monastic establishment in Europe, in particular because of the survival of its extensive water-management system of reservoirs and channels".

The monastery is a former Cistercian abate founded in 1138 and finally settled in Maulbronn in 1147. The monastery developed to an economic, social and political center of the region. The dimensions of the complex give evidence of that.

From 1156, the monastery became a bailiwick of the Holy Roman Empire by Emperor Friedrich II (Barbarossa). The responsibilities varied during times. In 1525, during the German Peasants' War, the monastery averted disaster of demolition. After secularization in 1534, Christoph, Duke of Wuerttemberg, established a Protestant seminary in 1556, which still exists.

The building complex consists of church and monastery, the cloister, a large monastery yard surrounded by a gate tower, a gate chapel (for women), a hostel for pilgrims, accommodation for priests, a mill, a coopery, a large store house and other economic buildings. It was here – in the monastery's narthex, called "the Paradise" – that Gothic design was first implemented in the German-speaking world.

The complex, surrounded by turreted, stony walls and a tower gate, today houses the Maulbronn town hall and other administrative offices, several restaurants, an Evangelical seminary in the Wuerttemberg tradition and a boarding school.

The buildings (selection)

The building ensemble offers a homogenous general view, not least because of the used regional sandstones, but includes also outstanding architectonical highlights. The main entrance with the iron-mounting door flanked by sandstone columns stands for the pure Romanesque style in the 12th century.

Paradise (Paradies)

The entrance hall to the monastery church, called *Paradies*, is a masterpiece of an unknown master builder. The Early Gothic designed hall with typical cross-ribbed vaults and individually arranged Romanic pilasters and early Gothic style elements from northern France is stone-faced today. Preserved remnants of color dating back to around 1430 indicate that the hall was painted.

Monastery Church (Klosterkirche)

This church still reflects the simple Romanesque design of the Cistercian order despite the Gothic transformation. The Romanesque nave was altered in the 15th century by adding a Gothic net vault and Gothic founder chapels.

The central sandstone crucifix in front of the choir screen is an important late Gothic sculpturing artwork by the master mason Conrad von Sinsheim in 1473. He carved the crucifix and Christ's body out of a single block. There are many big and small, visible and hidden stone artworks, sculptures and sculptured pieces not only inside the church and in the other monastery buildings. Worth to note are the cloister windows, the high medieval corbels adorned with plant and animal motifs in the west wing of the cloister, the leave capitals in the Paradise and the Monks Refectory.

Fountain house (Brunnenhaus)

The stony part of the circular shaped *Brunnenhaus* with five tracery windows was built in the 13th and 14th centuries. The lower bowl as well as

Figure 1.9.3 Minster built of yellow and reddish Maulbronn Sandstone.

the bronze top remained from the medieval fountain. The other parts had been added as new pieces in the 19th century.

Building stones

One quarry ("*Klostersteinbruch am Schafhof*", Figure 1.9.4) belonged to the foundation setting of the monastery. Thus, right from the start since 1170 the building material was the yellow to brownish Maulbronn Sandstone (see Chapter 2.4), from 1350 onward other quarries with reddish sandstones were used (Burggraff 2013). The sandstones had been first exploited and treated by the Cistercian monks for all kinds of use: rough stones, cut stones, ashlars, stone masonries and sculptures. Nearly all buildings are stone-faced and thus give a homogenous picture as well as a good impression of the stone varieties and their weathering features.

Most of the monastery buildings had been built around 1170, the farm buildings including mill and magazine up to 1250. The quarry was mined up to 1350. The building activity was low in the following 500 years and the demand for stones was little. Around 1850, in connection with the sale of the monastery, many buildings, bridges, roads and moats had been renovated or replaced by new buildings. Sandstone quarries opened and reopened, and quarrymen and craftsmen settled in Maulbronn. At the end of the 19th century, some 350 to 400 men were working in the quarries and factories (Ehlers 2013).

The sandstones are notably weathering when they are exposed to humidity, infiltrated with salt or when they are placed flatwise within the building. They tend to sand and scale, and they are often covered by moss and lichen. Considering the age of the buildings, the sandstones are relatively moderately damaged.

Restoration

All through the time, the sandstones had been repaired and replaced, if necessary. A comprehensive restoration on the façades and buttress of the *Klosterkirche* took place in 2010–2012 (Hollerung Restaurierung GmbH 2013). Besides the remediation of damages caused by construction defects, the sandstones had been cleaned and treated with fungicides. Some of them had to be replaced by new stone blocks or had been repaired with adapted acrylate mortar or by inserting stone pieces. Most of the historic and modern restoration measures are nearly invisible because of the permanent use of the regional sandstones.

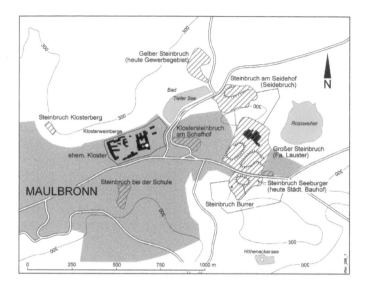

Figure 1.9.4 Map illustrating the location and extent of the Maulbronn monastery, the village (grey), the sandstone quarries (hatched in red) and the lakes.

Source: Werner et al. 2013.

1.10   Regensburg Old Town

Klaus Poschlod

Introduction

The outstanding universal value of the "Old Town of Regensburg with *Stadtamhof*" warranted inscription on the World Heritage List in 2006. This selection identifies the historic medieval town as being unique to the world. The complete Heritage Site in its entirety extends over 183 hectares and comprises almost 1,000 individual, historic buildings. This corresponds to the expansion of the city in 1320.

Regensburg's architecture reflects the role of the city as a medieval trading center and its influence on the region north of the Alps. Regensburg was a significant emporium alongside the continental trading routes to Italy, Bohemia, Russia and Byzantium. In addition, the city had numerous links to the intercontinental silk routes. This enabled a substantial

Figure 1.10.1 Stone Bridge and Cathedral (2009, before last restoration).

exchange of cultural and architectural influences, which still constitute the distinctive townscape of the city.

The Old Town of Regensburg displays an extraordinary testimony of cultural traditions of the Holy Roman Empire. In the High Middle Ages, Regensburg was favored as the meeting place for Imperial Assemblies, whereas, in more recent European history, it likewise served as the seat of the Perpetual Imperial Diet from 1663 to 1806. The remains of two royal palaces (*Kaiserpfalz*), which date from the 9th century, and the numerous well-preserved historical buildings are evidence of the former wealth and political significance of the city.

The Old Town of Regensburg is an excellent example of an inner-European medieval trading town of which the different historical stages of development are well preserved. The buildings in particular illustrate the history of trade from the 11th up to the 14th century with Roman, Romanesque and Gothic elements. They still shape the cityscape with high buildings, dark, narrow streets and strong fortifications. This includes patrician houses, residential towers, many churches and monasteries as well as the Stone Bridge from the 12th century.

With the decline of the commercial metropolis in the Late Middle Ages, the building activities nearly came to a standstill and large parts of the medieval city complex were preserved in an exceptional wholeness. The Old Town was also largely spared from destruction during World War II.

History

The Celtic name Radasbona has been passed on as the oldest name of a prehistoric settlement. A Roman cohort fort was built in the area of the present district of Kumpfmühl around AD 90. In AD 179, during the reign of the Emperor Marc Aurel, the Roman legionary camp Castra Regina (camp at the river Regen) of the Legio III Italica was founded.

In the 6th century, Castra Regina, now called Reganespurc, became ducal residence of the Agilolfinger and the first Bavarian capital. The Holy Bonifatius instituted the diocese of Regensburg in 739. In 788, Charlemagne dismissed the Grand Duke Tassilo III, the last Agilolfinger. During the 11th and 12th centuries, the crusader army gathered in Regensburg and three times left for the Holy Land. The economic boom during the 12th and 13th centuries, including long-distance trade to Paris, Venice and Kiev, transformed Regensburg to one of the wealthiest and most populous cities at that time.

Between 1135 and 1146, the Stone Bridge was built. This medieval building served as a model for many other bridge constructions such as the Charles Bridge in Prague.

The Emperor Barbarossa disposed Henry the Lion as Duke of Bavaria at the Imperial Assembly in Regensburg in 1180. The Wittelsbachs became the ruling family in Bavaria. The Emperor Frederick II conferred on Regensburg the right of self-government in 1245 by the privilege of "having a mayor and a council" (Regensburg remained a free imperial city until 1803).

In succession to a Romanesque cathedral of which a tower (*Eselsturm*) is still preserved, the construction of the Gothic cathedral began around 1275. Since 1450, the building was roofed and could be used. (The final assembly of the two cathedral towers and the spires did not take place until 1859 to 1869).

The "Holy Roman Empire of the German Nation" was dissolved in Regensburg in 1806. In 1809, Regensburg was bombarded and stormed by French troops under command of Napoleon. In 1810, Napoleon forced the Electorate Chancellor Dalberg to cede the spiritual principality of Regensburg to the newly formed Kingdom of Bavaria. Regensburg became the capital of the newly formed Bavarian district Regen and in

Figure 1.10.2 Cathedral of Regensburg (2019).

1838, the capital of the district Oberpfalz and Regensburg, which essentially corresponds to the boundaries of today's administrative district Oberpfalz.

The railway lines to Munich and Nuremberg were opened in 1859. The opening of the Luitpold Harbour (today: *Westhafen*) took place in 1910. In 1938, the synagogue was destroyed during the so-called *Reichskristallnacht*. In 1945, the Danube bridges were blown up. Women demonstrated for a peaceful surrender of the city.

Natural stones

The cathedral, which has been constructed since 1275, consists mainly of two stones. One is a Jurassic massive limestone (so-called *Donaukalkstein*) mined in various quarries along the Danube River and also shipped via the Danube. The other stone used is the Regensburg Green Sandstone (see Chapter 2.7). During the presence of the cathedral, some of the original stones have been replaced by foreign rocks, for example limestones from the vicinity of Pula (Istria/Croatia). Parts of the spires were recreated in concrete for safety reasons. The Stone Bridge mainly consists of the same regional stones (*Donaukalkstein* and *Regensburger Grünsandstein*) (Poschlod et al. 2017) whereas here – albeit to a small extent – further stones were used: carbonate-free sandstones of the *Schutzfels-Formation* and the stratigraphical younger marly limestones of the *Pulverturm-beds* (both of Cretaceous age) that outcrop at the southern edge of Regensburg. During the restoration of the Stone Bridge, which was completed in 2019, the intermediate concrete balustrade was replaced by a granite balustrade. The road surface was also renewed using granites from Flossenbürg and Thanstein, located in the Oberpfalz (approx. 100 km north of Regensburg).

The preserved Roman buildings in Regensburg also consist of the Jurassic limestones (*Donaukalksteine*) from the quarries at the Danube. The best-known is the one near Kapfelberg where traces of Roman mining are still to be found today (Holzhauser 2011). The quarry is located between Bad Abbach and Kelheim on the northern bank of the Danube. It was in operation for over 1,800 years. It experienced its prime during Roman times and the Middle Ages and has been used for many other buildings in Regensburg.

Besides these sedimentary rocks, other types of rocks were used as well, namely granites. The so-called residential towers (such as the "*Römer-Turm*" near the cathedral) were often built from granites from the forest of Regensburg. These are porphyric, coarse-grained granites, the so-called crystal granites (*Kristall-Granit*).

Figure 1.10.3 Preserved wall of Castra Regina (2019).

Chapter 2

Natural stones at UNESCO sites in Germany

Contents

2.1   Weser Sandstone (*Wesersandstein*)

Jochen Lepper

Brief geological description

Weser Sandstone is the trade name for fluvial red beds comprising three different main types of workable sandstones:

> Laminated Red Weser Sandstone (RWS; *Roter Wesersandstein*)
> Massive Red Weser Sandstone (RWS; *Roter Wesersandstein*)
> Grey Weser Sandstone (GWS; *Grauer Wesersandstein*)

According to its feldspar content, RWS has to be classified as subarkosic wacke, in contrast to GWS representing a subarkose. All of these three varieties developed in different fluvial environments: GWS is considered to have been deposited in a braided river system mainly by downstream accretion and as channel deposits; the massive RWS as point bars in a meandering river system with lateral accretion and finally the laminated RWS as fine-bedded floodplain sandstones. Paleogeographically and hydro-dynamically controlled, these three facies interfinger laterally and show variable workable thicknesses (Weber & Lepper 2002).

Geological age

Mesozoic – Triassic – Early Triassic – Middle Buntsandstein – Solling Formation; 247.4 million years

Occurrence, quarrying and use

The workable sandstones are embedded in the Solling Formation, which represents the uppermost formation of a total of four that together comprise the Middle Buntsandstein in Germany. The occurrence stretches from the northern rim of the Mesozoic uplands near Hannover, to the South, up to the upper reaches of the river Fulda, close to the town of Fulda, and, from West to East, from the upper Weser River to the western foreland of the Harz Mountains. Within this wide area, there are several provinces where workable sandstones have been quarried. One of the most important mining regions is in the Buntsandstein uplands on both sides of the upper Weser River between the town Bodenwerder in the North and Kassel in the South. Situated on the flanks of a denuded anticlinal Buntsandstein structure, the outcropping sandstones have been easily exploited since historical times. In medieval times, the sandstone had been used predominantly for the masonry of prominent stone buildings (churches, palaces and castles) and for fine handcrafted sculptures. It is estimated that much more than 500 quarries had been operated all through the time, but not simultaneously. There are much more than a dozen important medieval churches – one of them is the prominent former monastery complex of Corvey (Chapter 1.3) – and a couple of castles in the area to be conserved as inherited cultural objects built with the local Weser Sandstone.

The different qualities of Weser Sandstone had been used in former times according to their suitability for various intended purposes as well as manual stone working. The homogenous massive RWS had been used both as a raw quarry stone and as ashlar for structural constructions since medieval times. Bedding and cleavage controlled the dimensions of the unworked stones, whereas oversized raw blocks were roughly dressed manually. The massive homogenous fabric favors the manual making of precisely dimensioned ashlars with rectangular dressed planar surfaces, profiled pedestals, round columns and capitals and so on. At least since the 13th century, the laminated and platy splitting variety of RWS has been used for roofing, instead of using fire-trapping organic materials on the narrow neighboring houses in medieval agglomerations. In contrast, the GWS was notorious for its tool-damaging abrasiveness due to its comparably high quartz cementation. Accordingly, this sandstone had been used less and predominantly reserved for work pieces exposed to weathering like pedestals and quoins.

Nowadays, seven enterprises exploit about 9 quarries, providing among others suitable sandstones as exchange material to restore damaged cultural objects built originally with Weser Sandstone. Using industrial stone sawing and cutting, grinding, molding cutter and mechanical surface finishing machinery, the massive homogenous Red and Grey Weser Sandstones are processed into a wide range of products according to the actual demand (Lepper & Ehling 2018a).

Petrographical characteristics

The RWS is predominantly fine-grained, containing quartz as detrital and authigenic components (60–80%), feldspars (mainly K-feldspar; 15–25%) as well as clay matrix (illite, kaolinite and chlorite) and larger, detrital mica grains (up to > 16%) oriented parallel to the bedding planes. Grain shape is subangular to subrounded; grain contacts are elongated to concave-convex. Additionally, the subarkosic wacke is characterized by moderate quartz cement contents of 2–8% (Weber & Lepper 2002).

The GWS, in contrast, is predominantly medium-grained, containing quartz as detrital and authigenic components (65–85%), feldspars (mainly K-feldspar; 15–28%), a minor portion of clay matrix (illite and kaolinite) and larger detrital mica grains (both 2–12%). Grain shape is subangular to subrounded, grain contacts are mostly point-shaped, elongated and subordinately concave-convex. Additionally, the subarkoses are characterized by comparably high quartz cement content of 13–18% (Weber & Lepper 2002).

Technical parameters

Table 2.1.1 shows the diversity of the technical parameters. The GWS is qualified by far by high strengths combined with low porosities and water absorption, whereas the RWSs are characterized in general by lower strengths and relatively higher porosities and corresponding water absorptions.

Weathering resistance

The weathering phenomena and resistance are mainly controlled by the amount of quartz cement: a high amount of quartz cement and a low amount of clay minerals correlate with high density and low porosity and generally good petrophysical parameters, whereas minor quartz cement portions and relatively high amounts of clay matrix correlate with an augmented vulnerability to weathering. The main weathering phenomena are loss of material by sandy disaggregation, delamination, as well as scaling, spherical or lenticular cavities of only few millimeters in diameter resulting from dislodged non-cemented sand nodules.

Table 2.1.1 Technical Parameters of Weser Sandstones

	Laminated RWS	Massive RWS	GWS
Compressive strength [MPa]	46–110	37–116 (140)	99–180
Flexural strength (⊥) [MPa]	10–19	6–19 (21)	12–33
Water absorption (atm) [wt.%]	2.8–4.7	2.7–5.8	2.0–2.9
Porosity (eff) [%]	11–16	10–19	4–11

Source: Ehling and Lepper (2018a).

Figure 2.1.1 Highway bridge Arens-
burg (1939) made with
massive RWS.
Source: Lepper and Ehling (2018a).

Figure 2.1.2 Laminated RWS as façade
panels on the former
school in Arholzen.
Source: Lepper and Ehling (2018a).

Figure 2.1.3 Romanesque font (13th
century) with the baptism
of Jesus made of GWS.
Source: Lepper and Ehling (2018a).

Figure 2.1.4 Pavement with RWS and
GWS in Bad Karlshafen.
Source: Lepper and Ehling (2018a).

2.2   Nebra Sandstone (*Nebraer Sandstein*)

Angela Ehling

Brief geological description

The Nebra Sandstone belongs to the German Buntsandstein looking like it is typical with red, white, yellow and salmon pink colors. It occurs in thick beds, often with typical cross-bedding.

During the Buntsandstein period, the Central German region was controlled by the German Basin – an epicontinental area of sedimentation. Sandy material from the southern Bohemian Massif had been sedimented, mainly by rivers and in lacustrine facies. Thus, the sandstones show a big structural diversity with alternating sandy and clayey layers, cross-bedding, changes in grain size, fabric, strength and often showing drying cracks and lenticules. The dominating red color of the sandstones is caused by the arid climate during sedimentation.

Geological age

Mesozoic – Triassic – Early Triassic – Middle Buntsandstein – Hardegsen and Solling Formation; 248–245 million years.

Occurrence, quarrying and use

The Nebra Sandstone is part of the Triassic cuesta landscape in the South of Saxony-Anhalt, the so-called *Burgenlandschaft* (landscape with castles). Workable sandstones of the Early and Middle Buntsandstein occur there at several places. The Nebra Sandstone is the most famous one among them. It is forming a NW-SE running plateau from Allstedt and Lodersleben in the North up to Bad Bibra in the South. This plateau is cut by the river Unstrut in the southern part with the small city Nebra at the slope. Especially in this region, the Buntsandstein series is well exposed at both sides along the river.

The sediments of the Middle Buntsandstein are very thick. The Hardegsen Formation is about 75 m thick and contains four massive sandstone horizons. The Solling Formation starts with the Solling-Basissandstein (14 m) and ends with the so-called Chirotherien-Sandstein (-15 m), divided by 2–5 m thick clayey layers (Wehry 2005).

The sandstone beds are usually between 1 and 3 m. Clefts are usually wide-spaced up to 5 m. Therefore, it was and is possible to dig out relatively big blocks. The overburden was usually low.

Already in medieval times, the region was relatively densely populated and of particular importance for the development of the German empire. This fact and the exposed sandstone slopes along the river Unstrut with easy mining conditions favored a very early quarrying of the sandstones.

About 70 historical quarries are known from this area. Most of them were located at the river slopes near Nebra or at the slope of the Buntsandstein plateau north of Nebra.

Over centuries, these quarries had been used to cover the regional demand. The foundation of a stonecutter guild in the middle of the 18th century gives evidence of more systematic quarrying. In Lodersleben, mining took place also in the underground, following the inclining strong beds. The most important period of quarrying the Nebra Sandstone in the modern age began. It ended with WW I with only few quarries remaining. Between 1960 and 1997, no quarry was mined. In connection with the restoration of the Old National Gallery in Berlin, which was nearly completely faced with Nebra Sandstone, two quarries had been reactivated. One of them in Nebra-Kriebsholz is still open and offers the sandstone on demand.

The sandstones were used for all purposes, such as building stones, sculptures, gravestones, columns, cornices, millstones. At least since the 10th century, the sandstone has been used for walls, for the monasteries, castles and churches around, especially in Nebra and Memleben. There is proof that already in Romanesque times, the sandstone was used for churches in Merseburg, Naumburg and Magdeburg, in Gothic times even in Halle and other cities along the rivers within 100 km. All representative buildings in the *Burgenlandschaft* had been built with Nebra Sandstone. In the town of Nebra, sandstone was used for nearly all buildings.

The sandstone gained national importance at the end of the 18th century, when the river Unstrut had become a streamway. It was used for many bridges and watergates, and it could be transported to the north of Germany, especially to Berlin (Old National Gallery, 1876; see Chapter 1.2; columns at the famous "Red Town Hall", 1871).

At the end of the 19th century, when the railway became part of the transportation system, the Nebra Sandstone was used even for several public buildings in the northern cities by the sea such as Hamburg, Rostock and Greifswald.

After World War II, the demand on natural stones decreased and there are only few examples for the use of Nebra Sandstone in the 1950s in some East German cities.

Since the reactivation of two quarries initiated by the restoration of the Old National Gallery in Berlin, the Nebra Sandstone is still on the market. The variation in terms of color and structure is the reason that architects do not like to use the sandstone in modern architecture. It is mostly used for restoration today.

Petrographical characteristics

The sandstone is homogenous, but mostly cross-bedded. The colors vary extremely: from red salmon pink to grey, white, and yellow. It is fine- to medium-grained with an average grain size of 0.15–0.28 mm (0.12–0.42 mm).

The Nebra Sandstone is dominantly siliceous-bound. Besides the moderately developed grain contact intensity, diagenetic quartz growth exists. This is the base for the good building sandstone quality. Nevertheless, there are parts with clayey and/or carbonate pore fillings additional to or substituting the siliceous cement.

Composition: Quartz 62–90%, K-feldspar (+ little plagioclase) 7–25%, clay minerals (smectite, illite, seldom kaolinite) 1–3%, accessory minerals 1–2%, intermittent carbonates.

Technical parameters

According to the varying fabric, the technical parameters vary, too.

Pores are inhomogeneously distributed between the layers and within them. Average data of the building sandstone quality are as follows: compressive strength: 41–69 MPa; flexural strength ($\perp$): 3–5 MPa; water absorption (atm): 68 wt.%; and porosity (eff): 18–28%.

Weathering resistance

The weathering resistance of the Nebra Sandstone depends on the intensity of the siliceous bond as well as the occurrence of clayey layers,

Figure 2.2.1 Steinernes Festbuch (Stony festival book) Großjena (1722), detail with Duke Christian within the bedrock of Buntsandstein.

Figure 2.2.2 Nebra Sandstone panel with sedimentary structures on the Old National Gallery in Berlin.

Source: Ehling and Siedel (2011).

Figures 2.2.3 Baroque tombstones at the church St. Pankratius in Lodersleben, Saxony-Anhalt.

lenses and calcareous ingredients. Besides well-resistant sandstones, others reveal texture-following back weathering and loss of weak lenses and particles. Rounding of the edges, development of black crusts and their scaling are typical weathering phenomena. Delamination is typical when the sandstones are placed perpendicular to the stratification (Figure 2.2.3).

2.3 Kylltal Sandstone (Kylltaler Sandstein)

Friedrich Häfner

Brief geological description

The Kylltal Sandstone belongs to the German Buntsandstein. In the southern Eifel Mountains, geologically situated in the western part of the Middle-European Basin, claystones, siltstones, sandstones and conglomerates were deposited in an arid to semiarid climate during Early Triassic – the German Buntsandstein.

The sedimentation took place in a braided river system. The Bunt-sandstein stratigraphy of the southern Eifel Mts. is shown in Figure 2.3.1. The whole sequence has a thickness of about 300 m.

Geological age

Mesozoic – Triassic – Early Triassic – Upper Buntsandstein – Voltzien-Sandstein; 251–243 million years.

Occurrence, quarrying and use

The Buntsandstein beds roughly cover an area of about 40 x 5 km in the southern Eifel. Their maximum extension they reach north of Kyllburg around the valley of the river Kyll. The Romans already extracted copper

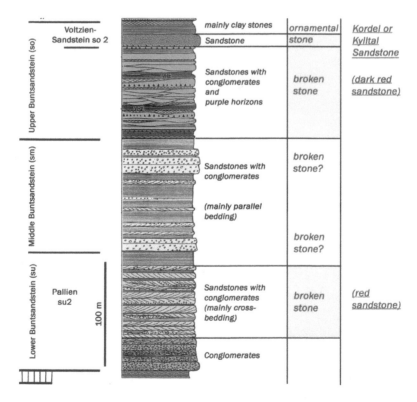

Figure 2.3.1 Buntsandstein in the southern Eifel Mts. with the frequently used natural stones.

Source: Modified after Wagner et al. 2012.

ore near Butzweiler west of the Kyll Valley. The copper mineralization with malachite and azurite was bound to occasionally occurring Carboniferous wooden relicts in the sandy sediments. Later on, instead of copper ore, sandstones were extracted to support the Romans construction activities in Trier (Figure 2.3.2).

At that time, Trier was a capital of the Roman Empire. In the 19th up to the early 20th century, many sandstone quarries existed in the Kyll Valley and its surroundings. Villages like Aach, Butzweiler, Malberg, Neidenbach, Neuheilenbach and Kordel were hotspots of the production of building and ornamental stones (Figure 2.3.3). The multitude of villages with quarries is reflected by several trade names, as there are *Kyllburger Sandstein*, *Kylltaler Sandstein* and *Kordeler Sandstein*. In 1996, 11 quarries were still active (Beyer 1996); in 2020, only three of them are still in operation. Only one or two thick beds are excavated by blasting in the quarries.

Figure 2.3.2 Roman copper mine and sandstone quarry *"Pützlöcher"*. Cultural monument near Butzweiler (Kylltal).

Source: Photo by Grubert.

Figure 2.3.3 View of the sandstone quarry in the Kylltal near Kordel, which is temporarily out of order.

Source: Photo M. Auras.

The Kylltal Sandstone has been used for several applications since Roman times, for example, as huge blocks (Porta Nigra in Trier, see Chapter 1.8, Figure 1.8.1), ashlars, columns (Cathedral in Trier, Chapter 1.8), for reliefs (*Igeler Säule* near Trier) and sculpturing. Further historic examples are the church St. Amandus (Kordel), the Abbey of Clervaux (Luxemburg), the Cathedral of Strasbourg (France) and the Prince Palace in Wittlich (Eifel).

Modern examples are the University of Trier, the town hall of Bitburg, the façade of the craftsmen's school in Kaiserslautern, a crucifixion group in the cemetery of St. Peter in Frankfurt a. M. and an office building in Koblenz.

Modern usage includes masonry, dry stonewalls, cladding of retaining walls, window and door frames, floor and façade tiles, staircases, balusters, fountains, tombstones and sculptures.

Petrographical characteristics

The sandstones are of white, grey, yellowish or red color. They are fine- to middle-grained; occasionally clay lenses and green plots occur. Parallel-bedded muscovite is enriched at bedding planes. The sandstones are mostly homogeneous, well sorted with small pores (Beyer 1996). They are classified as lithic subarkose.

The sandstones are composed of 75% components, 14% cement, (argillaceous, ferritic, rarely carbonate) and a pore volume of 11%. The components consist of quartz, 48%; lithoclasts, 37%; feldspar, 8%; mica, 5%; and carbonate, 2% (Grimm 2018). Beyer (1996) published the following total mineral composition of the Kylltal sandstones in Vol.-% (rounded): quartz, 40–51%; feldspar, 6–16%; lithoclasts, 7–18%; opaque minerals, 0.3–5%; mica (muscovite and biotite), 0–3.5%; and authigenic minerals (cement), 4–21%.

Technical parameters

Beyer (1996) examined 26 samples from 12 quarries in the Kyll Valley for analyzing the petrographical and technical properties of the sandstones. The minimum size of raw blocks demanded by Beyer (1996, by citing Singewald 1992) should be 0.4 m³. Nearly 90% of the examined quarries complied with this requirement. All samples exceeded the claimed water absorption of less than 0.5 wt.%, and all of them had a compressive strength of more than 30 MPa. Twenty-two out of 26 samples were resistant in a freeze-thaw-test with ten cycles. The most important results are summarized in Table 2.3.1.

Figure 2.3.4 Specimen of greyish white sandstone from Kordel (Kylltal).

Source: Photo by Grubert.

Figure 2.3.5 Sample on approval of *Kylltaler Sandstein* GmbH, sanded, format 20 × 30 cm (Malberg quarry).

Table 2.3.1 Technical Parameters of the Kylltal Sandstones

Compressive Strength [MPa]	Flexural Strength (⊥) [MPa]	Porosity (eff) [Vol.-%]	Water Absorption (atm) [wt.%]
49–97	5.5–8.0	9.99–20.37	3.5–7.6

Source: Beyer 1996; Grimm 2018.

Weathering resistance

The weathering resistance can be indicated as generally good – depending on the quarry and the beds the blocks had been extracted from. Sometimes lenses of clay are efflorescing; the sandstones show sanding and exfoliation.

2.4 Schilfsandstein in SW-Germany – Maulbronn Sandstone (*Maulbronner Sandstein*)

Angela Ehling and Wolfgang Werner

Brief geological description

The German Schilfsandstein occurs in many regions with Triassic rocks: in the Lower Saxonian Mountains region, in the Thuringian Basin, but the largest surficial extension can be found in SouthWest-Germany, in the so-called South German cuesta landscape. Due to the main localities with numerous quarries, the sandstones of the Schilfsandstein are known there under different names: The most important ones are *Maulbronn, Weil, Mühlbach, Niederhofen, Pfaffenhofen, Freudental, Heilbronn, Stuttgart, Renfrizhausen* in Baden-Württemberg as well as *Sand* in the North of Bavaria.

The typical German Schilfsandstein is a finely laminated, fine-grained, grain- and clayey-ferritic bound arkose with mainly brown-reddish, violet-reddish, yellowish, and greenish colors. These colors vary closely spaced, thus leading to spotted, stripped or flamed structures. Occasionally, plant remnants occur within the sandstones; its German name is derived from these plants (*Schilf* = reed), which, however, are remnants of large equiseta. Total thickness of the Schilfsandstein as well as the amount of clay and bed thickness vary at short distances. These variations are caused by the sedimentation of fine sands in N-S and NE-SW running channels surrounded by areas of clay sedimentation. The sand was transported from Northern Europe into a shallow sea (Figure 2.4.1).

The Maulbronn Sandstone is the main building material, which has been used at the UNESCO Site of the monastery in Maulbronn (see Chapter 1.9). The sandstone occurs in two main varieties: a homogenous yellow-brownish and reddish-brownish, dark-brown veined sandstone and a brown-reddish flamed sandstone. Trade names are also *Favorit-* and *Kosak-Sandstein*.

Geological age

Mesozoic – Triassic – Late Triassic (German Middle Keuper) – Stuttgart Formation (Schilfsandstein); 231–230 million years.

Occurrence, quarrying and use

The Maulbronn Sandstone occurs directly in the city of Maulbronn. The quarries have been situated at its northern and eastern periphery (Figure 1.9.1). Even in the neighboring cities of Knittlingen, Schmie and Freudenstein, Schilfsandstein had been mined in the past.

The sandstone has a total thickness of 15 to 20 m in the quarries. It is mostly flat bedded. The sandstone beds are usually between 1.2 and 2.5 m thick. Clefts are mostly wide-spaced. Therefore, it was and is possible to dig out relatively big blocks.

The brown-reddish flamed sandstone is still quarried at the eastern periphery of Maulbronn. Due to its good workability, the sandstone has

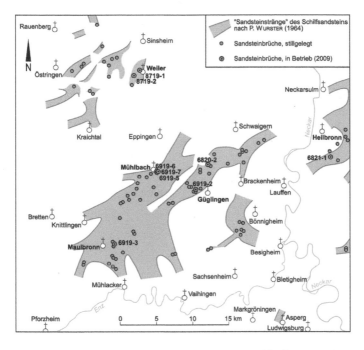

Figure 2.4.1 Distribution of Schilfsandstein in channel fills. Abandoned and still producing quarries (the latter marked with numbers) (LGRB).

Figure 2.4.2 Quarry wall in the large quarry of Maulbronn with a 15 m thick Schilfsandstein bed.

been used for all possible kinds of sandstone use – for building stones, as well as for sculptures and (in modern times) for façade tiles.

The comprehensive building ensemble of the Cistercian monastery and other historical buildings in Maulbronn and the neighboring cities give evidence for the manifold use of the sandstone since more than 900 years. Other important historical buildings are the State Theatre, the New Palace, the Queen Olga Building and the catholic Cathedral St. Eberhard in Stuttgart, as well as the *Erbprinzenpalais* (today Federal Supreme Court) in Karlsruhe. Example for a modern building is the Archiepiscopal Archive in Freiburg i. Br. (Werner et al. 2013). By the way, one of the German Schilfsandstein varieties, probably from Sand in Bavaria, had been used for a building of the Russian Trade Bank (1889) in St. Petersburg at Bolshaya Morskaya ul. 32 (Savchenok 2009).

Petrographical characteristics

The Maulbronn Sandstone occurs in two main varieties: the yellow-brownish (veined) and the brown-reddish flamed ones. A third variety with yellow-grey and greenish-yellow sandstones has been quarried in the past in the village of Schmie. The sandstone in Knittlingen shows yellow colors, the sandstone in Freudenstein is yellowish with dark brown veins and patches.

The sandstone is in general uniformly fine-grained with a mean grain size between 0.12 and 0.18 mm. It is characterized by a fine parallel bedding, seldom cross-bedding and brown-reddish crescent marks. The typical "flamed" image results from cutting these marks.

The average composition is as follows: quartz, 25–27%; lithoclasts, 25–49%; (K-) feldspar, 23–48%; mica, 2–4%; clay minerals (chlorite + illite), 6%; Fe-hydroxides and -oxides, 5–8%; and accessory minerals as there are carbonate, rutile, zircon, 1–2%.

The grains show mainly longitudinal and point contacts. Additionally, diagenetic thin quartz growth and often thick feldspar growth as well as quartz (rarely carbonate) cement, clayey-ferritic fringes and pore fillings compose a relative intensive bond and cause the good compressive strength of the sandstone.

Technical parameters

The technical parameters, especially for the compressive strength, reflect that they are of an average building sandstone quality.

Table 2.4.1 Technical Parameters of the Maulbronn Sandstone

Compressive Strength [MPa]	Flexural Strength (⊥) [MPa]	Water Absorption (atm) [wt.%]	Porosity (eff) [%]
64–83	6–10	6–7 (-14)	15–23

Weathering resistance

The Maulbronn Sandstone is a relatively durable sandstone as to be seen at the numerous historical buildings. Like all clayey bound sandstones, it is vulnerable to humidity and tends to sanding and scaling in case of moisture and salt attack, especially when placed flatwise within the building. Nevertheless, buildings made of this sandstone maintained over centuries.

Figure 2.4.3 The two main varieties of the Maulbronn Sandstone, the flamed reddish and the light brown types.

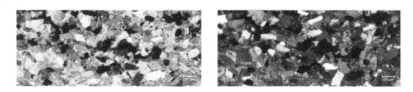

Figure 2.4.4 Thin section of a Maulbronn Sandstone (left – normal light and right – polarized).

2.5 Thuringian Rhaethian Sandstone (*Thüringer Rätsandstein*)

Lutz Katzschmann

Brief geological description

Sands and other fine clastic material of the Middle Rhaetian had been sedimented in a fluvial-deltaic environment, those of the Upper Rhaetian at an extensive shelf under shallow marine conditions. Small remnants of these Rhaetian sediments are preserved in trench areas of various fault zones within the Thuringian Basin. The Thuringian Rhaetian Sandstones are yellowish-white to ocher-colored, predominantly fine to medium grained, well-sorted quartz sandstones. Limonite precipitations (spots, Liesegang's rings) partially occur in the indistinctly stratified rocks. They were and are quarried from beds of the Middle and Upper Rhaetian.

Geological age

Mesozoic – Triassic – Upper Triassic – Rhaetian (German Upper Keuper) – Exter Formation; 208.5–201.3 million years.

Occurrence, quarrying and use

The classic mining areas are on top of the elongated Seeberg Mountain near Gotha, at the Rhönberg Mountain near Wandersleben and near Madelungen/Krauthausen (north of Eisenach) in the western Thuringian Basin (Hoppe 1939; Katzschmann et al. 2006).

The quarries at the Seeberg Mountain (e.g., *Kammerbruch* quarry) mine the middle Rhaetian Sandstones, which are composed of 0.50 to 2.50 m thick massive sandstone beds with intercalated clay/siltstone layers. The thickness of the sandstone beds decreases in the upper Rhaetian series. The quarry operators named individual characteristic beds depending on their properties and suitability (e.g., "*Wappen*" (coat of arms), "*Schleifstein*" (grindstone), "*Schäder*" (rotten stone)) (Klaua 1964).

The sandstones have been exploited since historical times at all locations mentioned earlier. Since the Middle Ages, the sandstones have been used primarily for masonry and decorative elements of magnificent buildings as there are castles, palaces and churches. It was not unusual to transport the stones over distances of more than 20 km. Two quarries are currently in operation on top of the Seeberg Mountain. The trade name of this sandstone is *Seeberg-Sandstein*.

Since the Middle Ages, the Raethian Sandstone has been used as raw quarry stone and ashlar as well as a sculptural stone and building stone (e.g., for columns, capitals, profiled pedestals, windowsills and frames). The best-known examples of its use include the UNESCO Site Wartburg near Eisenach (sandstones from the quarries near Krauthausen and from the Seeberg Mountain as replacement material; see Figure 2.5.3), the Erfurt Cathedral (*Seeberg-Sandstein, Rhönberg-Sandstein*, also Buntsandstein from the Tannroda anticline) and parts of the early medieval Mühlberg castle. Nowadays, the sandstones are used for all purposes of building construction, road and hydraulic engineering – as building stones, ashlar, curb stones, boundary stones, floor and wall slabs (outside, inside), steps, paving stone, sculpture stones, mill stone and grindstone or in open space design.

Petrographical characteristics

The Rhaethian Sandstones are well-sorted and mostly fine-grained sandstones. The detrital components consist of quartz, 90 – >95%; lithoclasts, up to 5%; K-feldspar, up to 2%; and mica, muscovite, 1%. Grain shape is subangular to subrounded, grain contacts are point, elongated and concave-convex. Additionally, the quartz sandstones (quartz arenite) are characterized by direct grain intergrowth and low cement content (quartz, limonite, kaolinite, illite).

Technical parameters

Table 2.5.1 Main Technical Parameters of Thuringian Rhaetian Sandstones

	Compressive Strength [MPa]	Flexural Strength (⊥) [MPa]	Water Absorption (atm) [wt.%]	Porosity (eff) [%]
Seeberg	58–110	6.3–12.2	3.9–7.4	12.3–15.8
Madelungen/ Krauthausen	64–90	3.6–6.5	5.4–6.4	21

Source: After Katzschmann & Lepper (o.J.), Grimm 2018; Katzschmann et al. 2006.

Weathering resistance

Depending on the intensity of silification, the sandstones are usually characterized by good petrophysical parameters (Table 2.5.1). This is the reason for the generally excellent weathering resistance. Dark to deep black rock surfaces are typical when the sandstones are directly exposed to rain (weather side). However, the less cemented stones show weathering phenomena such as sandy disaggregation, crumbling and incrustations in connection with stronger indirect moisture penetration, frost and/or salt exposure.

Other Rhaetian sandstones

Near-surface Rhaetian Sandstones also occur in other regions of Germany and were/are being quarried as building stones. These can also be described as white to yellow-brownish, mostly fine to medium, rarely coarse-grained quartz sandstones. They are often characterized by direct grain intergrowth, which leads to high strength and a good weathering resistance. They are also used for all purposes of building construction, but also as sculptural and paving stones as well as hydraulic engineering material.

The classic mining areas are in Haßberge (N Bavaria), around Tübingen (Württemberg), near Hildesheim, around Helmstedt and Velpke (all Lower Saxony) and in western Saxony-Anhalt. Besides the *Seeberg-Sandstein*, only one other Rhaetian Sandstone quarry near Buch in Haßberge (*Bucher Sandstein*) is still active.

According to Katzschmann & Lepper (o. J.), the Hildesheim Cathedral (*Hildesheimer Sandstein*), the historic Helmstedt town hall (*Velpker Sandstein*), Erlangen Castle (*Burgpreppacher Sandstein*), the church in Hamersleben (*Ummendorfer Sandstein*) and the Weser Renaissance castle Hämelschenburg (together with Schilfsandstein) were built with Rhaethian Sandstones.

Figure 2.5.1 Großer Seeberg, *Kammerbruch* quarry (10/2020).

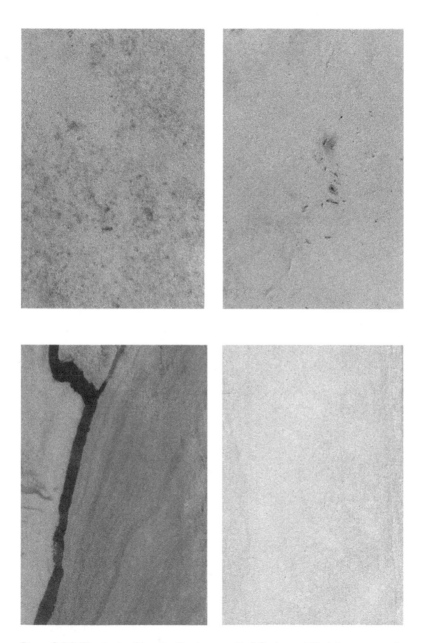

Figure 2.5.2 Thuringian Rhaetian Sandstones (1–3 Seeberg, 4 Madelungen, sample plates 10 × 15 cm).

Figure 2.5.3 World Heritage Site Wartburg castle with Palas (left, 12th century) near Eisenach.

Figure 2.5.4 Gravestone of the Rück-hold family in form of a tree trunk made of *Seeberg-Sandstein* (Historical Cemetery Weimar).

2.6 Obernkirchen Sandstone (*Obernkirchener Sandstein*)

Jochen Lepper

Brief geological description

The Obernkirchen Sandstone represents the most important variety within the group of the so-called Wealden sandstones in Germany. Together with the other varieties (from Rehburg, Deister, Süntel, Osterwald and Nesselberg), the Obernkirchen Sandstone was deposited in the proximal southern areas of the perimarine Lower Saxony Basin stretching from West to East between East-Holland and the Elbe River and the lower Weser River in the North and the Weser-Leine area in the northern portion of the Mesozoic uplands of Lower Saxony. A progradational fluvial system deposited in a distinct embayment situation a complex delta, and further toward the basin a barrier system composed of sand denuded from the uplands in the south. The Obernkirchen Sandstone is composed of very fine-grained mature barrier sandstones interbedded by some silty lagoon and swamp deposits, rich in carbonaceous matter and conserving two international attention-attracting horizons of dinosaur footprints (Figure 2.6.4). The quarried homogenous sandstone beds range from very fine-grained to coarse-silty. The color varies from light yellow grey to whitish grey. The total thickness of the useful beds is about 4 m; the thickness of the individually detached blocks varies by up to about one meter (Lepper & Ehling 2018b).

Geological age

Mesozoic – Cretaceous – Berriasium – Bückeberg Formation – Obernkirchen Member; 142 million years.

The traditional and in Germany well-known name "Wealden-Sand-stein" refers to the English Wealden beds, younger in age, but comprising a facies association similar to those basal beds of the Early Cretaceous in north-western Germany.

Occurrence, quarrying and use

The outcrop of the Obernkirchen Sandstone is restricted to the prominent ridge of the Bückeberge situated between Hannover and Minden. Other historical mining regions of similar Wealden sandstones are found further north in the Rehburg Mountains, in the Deister mountain range east of Hannover and further south in the Süntel, Nesselberg and Osterwald hills. All these mining regions were worked in historical times, whereas the Obernkirchen Sandstone is the only one won up to now on the Bücke-berge. There is a long chain of historical quarries following the belt of the outcropping, slightly inclined sandstones on the crestline here. Today, only one company is still exploiting these famous sandstones here. Covered by only few meters of Quaternary soft rocks, there are about 5 m useful rocks with bank thicknesses of 0.2 – < 1 m to be exploited, primarily formatted still on-site and subsequently processed in the workshops located in Obernkirchen at the traditional stone masons place.

The high quality of the sandstone was known already in historical times and reasoned its utilization: particularly for prominent sacral and profane buildings constructed not only nearby to the stone source but as well as further away. The Romanesque churches (the Cathedral of Minden and the Collegiate Church of Obernkirchen) both situated more or less nearby to these historical quarries bear witness to the use of Obernkirchen Sand-stone as a building stone since the 12th century. The building boom of the Renaissance period extensively used this sandstone as well: examples are, apart from a few town halls, situated in a limited distance from the

Bückeberge, the pompous frontage to the market-facing side of the prominent town hall in Bremen (Chapter 1.1) and even abroad the Royal palace
at Amsterdam, the town halls in Antwerp, Leiden and the Kronborg Castle
(DK). During 19th century, in St. Petersburg this sandstone has been used
for the Yusypova Palace, the bank building Nevskij Prospect 62 and the
railway station Novij Peterhof. Many more than 100 buildings in the Netherlands are reported to have been built by or with Obernkirchen Sandstone,
and in Denmark as well a formidable number of prominent castles, banks
and museums of different periods are listed (Lepper & Ehling 2018b).

The wide distribution of this sandstone far away from the extraction place
benefited from the transport facilities by ship downstream on the Weser
River. The Hanseatic harbor city of Bremen, situated on the riverbank of
the Weser, reserved the right to manage the trade with Obernkirchen Sandstone products, offering as well any further shipping by sea freight near
coast even to neighboring countries where the sandstone was well known
as "*Bremer Stein*". This sandstone has been used since early medieval times
not just as raw quarry stone but also for precisely dimensioned ashlars,
dressed round columns, capitals and pedestals. Notably during the Gothic
and Renaissance periods, fine tracery and delicate sculpture work, figural
as well as ornamental, were executed manually by skilled artisanal stonemasons. Nowadays, technical machinery has largely replaced the former
manual work. The modern portfolio comprises the full range of products
from simple raw blocks for gardening and landscaping, to dimensioned
building stones with different surface structures, paving stones of different
dimensions, sawn slabs for façade cladding, massive stair steps and bol-
lards, ornamental stones and, finally, to handcrafted sculptures.

Petrographical characteristics

The Obernkirchen Sandstone is a very fine-grained (mean: 0.05–0.09 mm),
well-sorted sandstone, locally embedding only very few middle-grained
lenses. The color, controlled by the measure of the Fe-hydroxide impregnations, ranges accordingly from white grey over light yellow to honey
yellow, sometimes overprinted by brownish concentric Fe-hydroxide
precipitation rings. Grain shape is edge-rounded to bad-rounded, due to
diagenetic induced quartz growth. The ratios of the components are detrital quartz, 83–90%; clay minerals (kaolinite, dickite), 7–12%; lithoclasts,
< 1%; feldspar, < 1–2%; limonite, generally < 0.5%; and accessories, 1–2%.

Technical parameters

Qualifying the Obernkirchen Sandstone is its distinct quartz cementation
controlling its petrophysical specifications: compressive strength, 64–107

Figure 2.6.1 Gable of the town hall in Bremen with Obernkirchen Sandstone.

Figure 2.6.2 Relief *"Löwenkampf"* (lions battle) at the Leibniz Building in Hannover (both Lepper & Ehling 2018b).

Figure 2.6.3 Quarry in Obernkirchen with bedrocks and detached sandstone blocks.

Figure 2.6.4 Iguanodon footprint preserved on a sandstone bedding plane in the quarry (length appr. 35 cm; Raddatz-Antusch 2019).

Figure 2.6.5 Thin section of the Obernkirchen Sandstone with mainly quartz with diagenetic rims and some dickite as pore fillings.

MPa; flexural strength, 7.3–9.8 MPa; water absorption, 5.1–5.3 wt.%; and porosity, 16.4–17.6 vol.% (Lepper & Ehling 2018b).

Weathering resistance

Due to the high quartz content combined with its quartz cementation, and both low water absorption and porosity, the weathering resistance has generally to be qualified as good to very good, reflected as well by its proven frost-thaw resistance as seen empirically on many non-affected historical buildings. The minor hygric expansion and low ratio of micropores favor this important quality additionally. According to variable moisture content in combination with the loading of de-icing salt, sanding and distinct loss of material, sometimes etching the bedding, can be observed. Long exposure over centuries may also lead to the development of a black surface layer as it is typical for many sandstones.

2.7 Regensburg Green Sandstone (*Regensburger Grünsandstein*)

Klaus Poschlod

Brief geological description

The North of Bavaria had been a continental landmass since the beginning of Cretaceous period up to earliest Late Cretaceous. The surface was exposed to weathering and erosion. The Early Cretaceous sand, clay, colored clays, bean ores and so on that had sedimentated during the Cretaceous transgression were preserved as the so-called Schutzfels Formation within karst holes of the underlying Jurassic limestones.

During Late Cretaceous (Upper Cenomanian), the sea expanded from the South to the mountain range of *Fränkische Alb* and thus, sand and marl, partly clay-, lime- and conglomerate beds had sedimentated in the former "Gulf of Regensburg". The oldest strata series of these sediments today build up the Regensburg Green Sandstone.

The Regensburg Formation is a varied sequence of different sedimentary rocks and thus, the formation is divided into an upper and a lower part. The roof rock is a clayey-silty marl, the *"Eibrunner Mergel"*. The bedrock is a Jurassic limestone and locally, within the karst holes, the previously mentioned sediments of the Schutzfels Formation.

Geological age

Mesozoic – Cretaceous – Late Cretaceous – Cenomanian – Regensburg Formation; 100–97 million years.

Occurrence, quarrying and use

The main occurrence of the Regensburg Green Sandstone is limited to a small region north, west, northwest and southwest of the city of Regensburg. Many quarries were situated near the river Danube in the region between Kelheim and Regensburg. The last mining activities took place around 1990 in a quarry west of Ihrlerstein. That is why the sandstone is called *Ihrlersteiner Grünsandstein*, too.

The Regensburg Green Sandstone had been quarried at least since Roman times. This is proved by the wall of Castra Regina, a Roman legion's camp (*Legionslager*), which was the birthplace of the town Regensburg (Dallmeier et al. 2004).

Because of its strong weathering, the stone was discredited during the reign of King Ludwig I in the middle of the 19th century. From 1880 on, it had not been used for new buildings for a long time. The real cause for the bad conservation status of the stones was the wrong selection of the sandstones within the quarries, because there are indeed weathering resistant beds but not all beds have a good quality. The types C1 and C2 are the most distributed varieties, followed by type B2 (explained next).

Regensburg and Munich are the cities, where the sandstones had been used most commonly. The portal of the Moosburg Cathedral (east of

Munich) is deemed to be the earliest medieval use of the Regensburg Green Sandstone around Munich. The earliest overall use of the sandstone in the city of Munich might be represented by the socle and the basin of the Otto-von-Wittelsbach-Fountain (circa 1610) in the Fountain Courtyard of the Residence. Besides the Residence, the Allerheiligenhof Church and the Old and New Pinakothek are the most important buildings in Munich made of Regensburg Green Sandstone. In Regensburg, the Old Town of which was designated as UNESCO World Heritage, there are many buildings made with this sandstone. The most important of these are the Regensburg Cathedral and the Stony Bridge, which is more than 800 years old (see Chapter 1.10). The sandstone can also be found at several buildings in other cities and villages along the river Donau, for example at the Liberation Hall (*Befreiungshalle*) in Kelheim (stairway) or the Lions Monument in Bad Abbach.

Petrographical characteristics

Intensely varying portions of sand and lime, different strength as well as varying sedimentary fabric in terms of grain size, color and fossil content characterize the Regensburg Green Sandstone. The varying intensity of the green coloration according to the glauconite content is very evident and played an important role regarding the selection and use of the sandstone.

The Regensburg Formation is divided into seven visually distinguishable rock types. Only two of them (C1 and C2) are appropriate to be used as building stones. Two other types (B2 and B4) are suitable for building stones only to a limited extent (Poschlod & Wamsler 2009). Type B2 is an olive-green to beige-grey, limy-marly fine sandstone with streaky fabric and sporadic fossils visible with the naked eye as well as a glauconite content higher than this of type B1. Type B4 is a light grey-brown limestone with

Table 2.7.1 Minimum and Maximum Values of Some Technical Parameters of the Types B2, B4, C1 and C2 of the Regensburg Green Sandstone (Poschlod & Wamsler 2009) and Unpublished Data of LfU-Laboratory (*)

	Type B2	Type B4	Type C1	Type C2
Compressive Strength [MPa]	22–168	42–160	37–124	13–46
Flexural Strength [MPa]	2–30	5–23.5	4–25	0.3–15
Water Absorption* (atm) [wt.%]	1.9–7.6	1.4–2.5	1.4–9.1	2.8–9.8
Porosity [vol.%]	3–22	1–11	3–14	5–28

Figures 2.7.1 Munich, Old and New Pinakothek.

Figure 2.7.3 Regensburg Green
Sandstone (15×20 cm).

Source: BGR collection.

Figure 2.7.2 Munich, Allerheiligenhof
Church.

Figure 2.7.4 Thin section Ihrlerstein_
untWS×20a — lower
shill-rich bed (Image
width 7.5 mm).

Source: Wilmsen & Niebuhr 2014.

a small sand fraction, a phacoidal-wavy fabric and typical thin iron oxide layers; glauconites are clearly visible. Type C1 is a light grey to beige-grey, mainly dolomitic bound, medium-grained sandstone; partly greenish, partly very porous, strong to very strong with scattered dark green glauconites. Type C2 is a grey-green, mainly calcitic bound medium-grained sandstone, moderately strong to strong with massive fabric, high glauconite content and numerous fossils such as white, thin-shelled 2–3 cm long, slightly bent fragments of shells, brachiopods and echinites.

A representative and exemplary composition of the sandstone is as follows: the components consist of quartz, 63%; glauconite, 15%; lithoclasts, 11%; fossil shells, 5%; K-feldspar, 3%; and accessories (chlorite, zircon, rutile, ore), 3%. They are surrounded and bound by a calcareous-micritic cementing material, which could amount up to 50% of the stone. The grain size averages out at 0.1–0.7 mm (Grimm 2018).

Technical parameters

These parameters underline the previously described petrographical variability.

Weathering resistance

The weathering resistance largely depends on the sandstone type, but overall, it is moderate. The sandstones exhibit all typical weathering phenomena sandstones can show, especially in frequently soaked zones, where the calcareous cement dissolves and the loss of particles increases.

2.8   Elbe Sandstone (Elbsandstein)

Heiner Siedel

Brief geological description

Elbe Sandstone is the general term for sandstone strata deposited in the shallow marine environment of a narrow strait between the Mid-European Island and the Western Sudetic Island in the northern part of the Bohemian Cretaceous Basin. The outcropping Elbe Sandstone beds

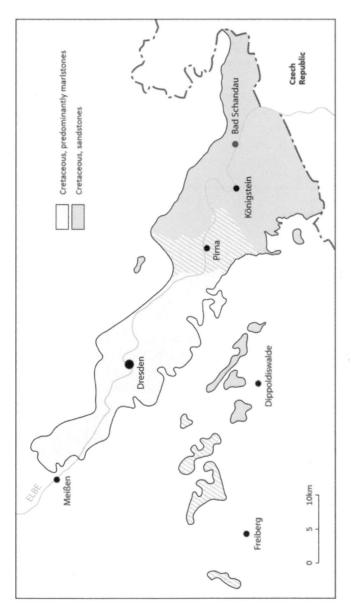

Figure 2.8.1 Sketch map of upper Cretaceous sediments along the Elbe Fault Zone, Saxony.

Source: Modified after Siedel et al. 2011.

follow the WNW oriented Elbe Fault Zone, although some smaller erosional remnants of Cretaceous sandstones (especially from Cenomanian, partially showing fluvial to estuarine lithofacies) are also located west of the fault zone on the crystalline basement of the Erzgebirge Mts. (Niederschöna/Grillenburg, Dippoldiswalde, Paulsdorf) (Figure 2.8.1). The latter were also quarried and can be attributed to the Elbe Sandstone deposits in a wider sense.

The Elbe Sandstone comprises approx. 400 m thick, quartz-rich, fine- to coarse-grained sediments with varying clay contents, rarely intercalated by thin clay strata, corresponding to sea-level changes during sedimentation. Due to later erosion, the sandstone forms a scenic area with cliffs and rock towers, the so-called *Sächsische Schweiz* (Saxon Switzerland).

Geological age

Mesozoic – Cretaceous – Late Cretaceous (upper Cenomanian to lower Coniacian). Most of the quarried sandstones are of Turonian to lower Coniacian age; 94–88 million years.

Occurrence, quarrying and use

The main area where outcropping Cretaceous sandstones have been quarried from the Middle Ages until today is situated between the town of Pirna (about 20 km southeast of Dresden) and the Czech border. This area stretches from Pirna to the southeast along the river Elbe with a maximum width of approx. 20 km. The first mention of sandstone quarrying in the Elbe Valley in historic documents dates from the 14th century. However, it must have commenced much earlier, since sandstones from this area were already used for the construction of the Meissen Cathedral, which started around 1250. Archaeological excavations of preceding buildings of the Residence Castle and the first stone bridge in Dresden witness the use of sandstones from the Elbe Valley in the early 13th century, maybe already in the late 12th century. The use of Cenomanian sandstones from Grillenburg and Niederschöna in the last third of the 12th century can be proven at architectural elements of the Altzella monastery (ca. 20 km west of Dresden).

Early quarrying in the Elbe Valley obviously started north of Pirna in the Lohmen region and developed southeastward, mainly along the river, over centuries. Quarries near by the river widened the steep Elbe Valley and provided good opportunities for a comfortable transport of sandstone blocks downstream by boats. The geographic situation of the quarries near the river Elbe facilitated a wide distribution to faraway places such as Torgau, Leipzig, Wittenberg, Berlin, Potsdam and Hamburg and even to Denmark already before the industrial era allowed transport by railway. At the peak of the development of the sandstone industry in the last third of the 19th century, nearly 400 quarries in the Elbe Sandstone produced different products for construction purposes. The number of quarries decreased significantly after the turn of the 19th century (Kutschke 2000). Nowadays, eight quarries still produce Elbe Sandstone for construction and restoration.

Due to its high mechanical strength the *Posta type* Elbe Sandstone has been mainly used as a building stone for structural components in construction, such as ashlar masonry and pillars, whereas the *Cotta type* Elbe Sandstone is preferred for sculpturing due to its fine grain size and lower mechanical strength (Figure 2.8.2). The latter has been utilized only since the late 15th century. Today, both the *Cotta* and the *Posta type* Elbe Sandstone are used for replacement of weathered stones on historic buildings and monuments as well as for modern buildings (façade cladding etc.) and for gardening and landscaping.

The builders of different objects belonging to World Cultural Heritage Sites in Germany used Elbe Sandstone as construction material. The Museum Island in Berlin is described in this book (Chapter 1.2). Another example is the castle of Potsdam-Sanssouci, belonging to the Palaces and Parks of Potsdam and Berlin UNESCO Site. It was planned by the architect Georg Wenzeslaus von Knobelsdorff and finished in 1747. The façade displays steps, pilaster strips, balustrades, door and window jambs, cornices and sculptures made of *Cotta type* Elbe sandstone (Figure 2.8.3). The Garden Kingdom of Dessau-Wörlitz comprises different historic gardens and palaces from the age of enlightenment in the Middle Elbe Region around the city of Dessau (Saxony-Anhalt). There, architectural elements made of Elbe Sandstone can be found on the façades of the palaces in Wörlitz, Oranienbaum and Mosigkau as well as on smaller buildings in the surrounding parks. Some of the Luther Memorials in Wittenberg, such as the Castle Church and the local church in Wittenberg, also present architectural elements of Elbe Sandstone on their facades, as well as different objects of the same material (tombstones etc.) in the interior. On the living house of Martin Luther in Wittenberg, a Renaissance portal from 1540 is situated (Figure 2.8.4). It is told that the wife of Luther, Katharina von Bora, had donated this portal of *Cotta type* Elbe Sandstone to her husband on the occasion of his 57th birthday, since it comprises a portrait of him at this age. The Mining Cultural Landscape Erzgebirge/Krušne hoři offers delicate examples for sculpturing with *Cotta type* sandstone by the altar

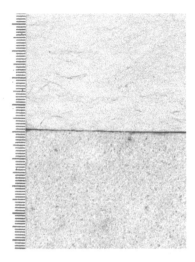

Figure 2.8.2 Macroscopic appearance of Elbe Sandstones: *Cotta* (above) and *Posta* type (below).

Figure 2.8.3 Sculptures made of *Cotta* type Elbe Sandstone on the façade of the Sanssouci Palace in Potsdam (1747).

Figure 2.8.4 Entrance portal of the living house of Martin Luther in Wittenberg (1540), made of *Cotta* type Elbe Sandstone.

Figure 2.8.5 The Baroque Church of Our Lady (*Frauenkirche*) in Dresden after reconstruction with *Posta* type Elbe Sandstone.

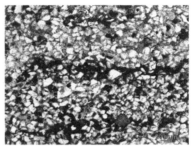

Figure 2.8.6 Microphotographs of *Posta* (left) and *Cotta type* sandstone (plane polarized light, pores impregnated with blue dye) (right).

and the epitaph of the Bünau family in the church of Lauenstein, which was the administrative center of the Altenberg mining area.

Moreover, Elbe Sandstone was used as building and sculpturing material for other well-known historic buildings in Germany, such as the Brandenburg Gate in Berlin, the Hartenfels Castle in Torgau or the Zwinger, the Opera House (*Semperoper*) and the Church of Our Lady (*Frauenkirche*, Figure 2.8.5) in Dresden. The latter was rebuilt from the ruin between 1994 and 2005 after destruction at the end of the Second World War, using original and new ashlars made of *Posta type* Elbe Sandstone. Private financial donations from all over Germany and other countries like the U.K. and the U.S.A. supported rebuilding. 15,000 m³ of fresh Elbe Sandstone were quarried for the reconstruction measures. The recent use of Elbe Sandstone for the reconstruction of the façades of the City Castles in Berlin and Potsdam is also based on its historic utilization on these buildings, which were destroyed during World War II.

Petrographical characteristics

The Elbe Sandstone used for construction can be generally classified as quartz arenite with > 90% quartz content in most cases. Traditionally, the stonemasons and builders distinguished two different types of Elbe Sandstone, the *Posta* and the *Cotta type* (Figure 2.8.6) according to petrography and technical properties. This practical classification is independent of stratigraphic features and was confirmed by comprehensive investigations of physical and technical properties (Grunert 1986).

The *Posta type* sandstone is a fine- to medium-, occasionally coarse-grained quartz arenite with very high amounts of quartz (in many cases nearly 100%). Accessories are kaolinite and feldspar. The grains are silica-cemented ore directly overgrown. Beside grey color, yellowish-brownish shades occur due to impregnation with accessory iron minerals such as limonite.

The *Cotta type* sandstone is a fine-grained quartz arenite, containing minor amounts of feldspar and clay minerals (mainly kaolinite, illite), in some cases also glauconite. Local accumulations of clay minerals are oriented ± parallel to bedding in flaser structures and may contain organic matter and limonite as well. Grains are silica-cemented or directly overgrown. Within clay accumulations, the respective minerals may form a pore-filling cement. As in *Posta type* sandstone, colors reach from grey to yellowish-brownish.

Technical parameters

Posta type sandstone reveals pore diameters in the range of some 10 μm up to some 100 μm, causing fast capillary suction and fast release of water while drying. It is frost-resistant and resistant to salt attack to a high extent. *Cotta type* sandstone has fine capillary pores (< 20 μm) and a distinct amount of micropores (< 0.1 μm). Its capillary suction as well as drying is much slower, compared to *Posta type* sandstone. Moreover, it is more vulnerable to salt attack (Siedel et al. 2011).

Weathering resistance

Posta type sandstone is mostly in a good condition even after several centuries of exposure as it is seen at historical buildings and monuments (Siedel 2013). Due to the mobilization and fixation of iron from the substrate

Table 2.8.1 Range and Average (in Brackets) of Technical Properties of Elbe Sandstone

	Compressive Strength (⊥) [MPa]	Flexural Strength (⊥) [MPa]	Water Absorption [wt.%]	Total Porosity [vol.%]
Posta type	36–76 (53)	2–8 (5)	5.2–11.2 (7.6)	17.8–25.1 (21.8)
Cotta type	20–57 (38)	2–6 (4)	7.1–14.0 (8.9)	19.2–27.4 (22.8)

Source: Grunert 1986.

at the surface in thin black layers ("patina"), it may turn black after some decades (depending on exposure situation). Extreme moisture and salt attack (e.g., at the base of buildings) may lead to granular disintegration (sanding). *Cotta type* sandstone, in contrast, more often shows weathering features like flaking, sanding and differential erosion after decades of exposure to moisture and salt attack. Bioturbated *Cotta type* sandstones may also show alveolar weathering in case of high salt load (Siedel 2010).

2.9 Colored "marbles" (*Buntmarmore*) in Germany

Thomas Kirnbauer and Gerda Schirrmeister

Introduction

Colored marine limestones of Silurian to Carboniferous age are widespread in the Rhenish Massif, the Harz Mountains and the northwestern parts of the Bohemian Massif (Germany; Figure 2.9.1). They were deposited in shelf and volcanic island positions of the Paleozoic terrane assemblages between the colliding super-continents Gondwana and Laurussia (Old Red Continent); the latter culminated in the Variscan Orogeny during the Upper Carboniferous. While the deposits in the Rhenish Massif and the Harz Mountains belong to the Rhenohercynian Zone of the Variscan orogen, the others are part of the Saxothuringian Zone.

These limestones were exploited in hundreds of quarries between the end of the 15th and the end of the 20th century. They were used as dimension stones mainly for interior decoration, locally for exterior work, too. Periods of high importance were the baroque, the Wilhelminian and the Nazi period, locally also the time after World War II. As early as the 18th century, they were exported to other countries, leading to worldwide use in the 20th century.

Due to their good polish, they were called "marbles"; for physical and mechanical properties see Table 2.9.1. The wide and unique range of colors, textures and structures of these fascinating technical marbles

Table 2.9.1 Physical and Mechanical Properties of German Colored "Marbles"

Compressive Strength [MPa]	Flexural Strength [MPa]	Water Absorption (atm) [wt. %]	Porosity (eff) [vol. %]
40–220	5–30 (max. 35)	0.02–1.1 (max. 2.5)	0.1–1.8 (max. 2.9)

stems from their sedimentological and diagenetic history, a weak Variscan metamorphism and, locally, from hydrothermal overprinting.

Silurian nodular limestones

During the Upper Silurian, marine limestones and limestone-shale turbidite sequences were deposited in the western part of the Saxothuringian Zone. Stratigraphically, the predominantly nodular limestones belong to the Ockerkalk Formation. The name refers to the ocher color, due to oxidation of pyrite to limonite by weathering. The colors of unweathered rocks vary from nearly black to grey and green. The limestones were quarried since the 18th century in Eastern Thuringia (near Saalfeld and Saalburg) and the neighboring Frankenwald region (near Ludwigstadt). Trading names were, for example, *Goldfleck* (Figure 2.9.2a) and *Saalburger Meergrün*. Prominent buildings with interior decoration are the Heidecksburg Castle in Rudolstadt (*Goldfleck*) and the Berlin Cathedral (*Meergrün*).

Middle to Upper Devonian reef limestones

During the Middle to Upper Devonian, a widespread carbonatic sedimentation and growth of reefs developed at the southern margin of the Old Red continent. Reef limestones from this "Devonian South Sea" are known from the Rhenish Massif and the Harz Mts. In the Eifel area, shallow water carbonates, deposited in a shelf lagoon environment, occur, while east of it, a carbonate platform with isolated shelf atolls and, near the shelf margin, carbonate complexes are found. Isolated but large carbonate complexes are bound to submarine volcanic rises and islands in the ocean basin in the south (Lahn region, Harz Mts.).

The most important quarry district with more than 100 quarries was the Lahn region in the southeastern Rhenish Massif. For approximately 400 years until the 70s of the 20th century, limestones were extracted and traded as Nassau Marble, since the 20th century as Lahn Marble (Kirnbauer 2008). They are used for decoration of the interior of countless buildings all over the world. The Empire State Building in New York, the Moscow Kremlin and the Tagore Castle in Kolkata are prominent examples. In Germany, thousands of churches, castles, palaces, public and office buildings, hotels and residential buildings are furnished with it, for example, the German UNESCO World Heritage Sites Würzburg Residence, Augustusburg and Falkenlust Palaces in Brühl and the cathedrals of Aachen, Cologne (see Chapter 1.4), Speyer and Trier (see Chapter 1.8). Important extracting and processing sites were situated in Balduinstein, Diez, Limburg, Schupbach, Wirbelau, Gaudernbach, Mudershausen and Villmar. Figures 2.9.2b–d show

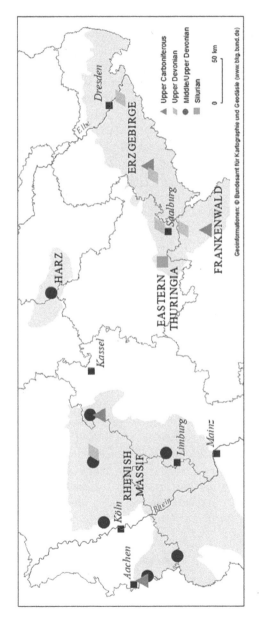

Figure 2.9.1 Sketch map of central Germany with regions of historic colored "marble" quarrying.

Source: Authors, Graphic design: J. Seltenheim.

three different varieties, representing main reef, fore reef and post reef facies. Similar reef limestones in association with submarine volcanism are known from the Elbingerode region in the Harz Mts. Reddish and black to grey limestones were exploited between the 16th and 19th centuries at the Krockstein quarries and at Hartenberg Hill. Examples of the Rübeland Marble can be found in churches in Potsdam.

The Sauerland region in the northeastern Rhenish Massif was of less economic importance. Despite individual examples from the 18th century, economically significant quarrying did not take place until beginning of industrialization in the second half of the 19th century. Important quarries and factories were situated near the reef complexes of Warstein, Brilon and Attendorn. Trading names of these Westphalian Marbles were, for example, *Alma, Elisabeth, Goldader, Mecklinghausen Grau* and *Lennestein*.

In the northeastern part of the Rhenish Massif, dark grey to black limestones were occasionally quarried as Waldeck Marbles (17th to the 19th century). In the northwestern part of the Rhenish Massif east of the river Rhine, colored marbles were used in the Düsseldorf (Ratingen, Neandertal) and Wuppertal area in the 17th and 20th centuries to a lesser extent. East of Cologne, colored and black reef limestones were extracted near Linde. An outstanding example is the Baroque Holy Stairs in Bonn. Quarrying ended in the 20th century (*Linder Schwarz*).

In the Eifel region west of the river Rhine, Devonian limestones are known from the Aachen region and areas further to the southeast. Predominantly dark limestones with isolated stromatoporoids are called *Aachener Blaustein*. It was used since Roman times; a prominent example is the medieval UNESCO World Heritage Site Aachen Cathedral. More colored limestones were mined as Eifel Marbles in the shelf lagoon environment further southeast. The use as "marble" is documented for Urft and Steinfeld (from the 17th to the 20th century), Eschweiler (19th century), Roderath (18th and 20th centuries), Neuenstein (19th century), Mülheim (around 1900), Üxheim (around 1900), Kerpen (from 1920 to 1950s, *Weinberg*) and Niederehe (1950s to 1980s, *Zisterzienser*).

Upper Devonian nodular limestones

While reef growth ceased in the mid-Frasnian, carbonate sedimentation persisted in the different marine basins between Gondwana and Laurussia during the entire Upper Devonian. Depending on the submarine relief, pelagic condensed limestones developed on intra-basinal rises, whereas shales and marls dominate the basinal sediments. Many limestones show the typical nodular structure due to diagenetic processes ("Kramenzelkalk").

In three East Thuringian areas, quarrying of nodular limestones was active between the end of the 19th and the end of the 20th century: north and south of Schleiz and near Saalfeld. The material was processed in Saalburg and therefore called Saalburg Marbles (Hartenstein & Lange 1991). The numerous grey and red varieties (some of them with cephalopods, Figure 2.9.2e) range from the typical nodular limestone *Fischersdorf* through the almost massive *Königsrot (Imperial)* to the regional metamorphic marble *Violet*. Saalburg Marbles can be found in many German towns, especially prominent in Berlin, for example, UNESCO Site Museum Island (see Chapter 1.2), Berlin Cathedral and Main Building of the Humboldt University.

In the Frankenwald region, nodular limestones were mined in more than 30 quarries since the 18th century. The most important deposits are located near Hof (*Theresienstein*), Kronach (*Wallenfels*) and Bad Steben (*Deutsch Rot*, Figure 2.9.2f). The quarrying ended in the early 1990s. Impressive examples for their usage are for instance the New Town Hall in Hannover (staircase, *Wallenfels*), the railway station building Lindau/Bodensee (*Theresienstein*) and the Bavarian National Museum in Munich (*Deutsch Rot*).

Upper Devonian limestones of Saxony from the deposits Grünau near Wildenfels (red spotted) and Maxen southeast of Dresden (grey and contact-metamorphic red varieties) were used regionally since the 16th century. The most prominent example is the Green Vault (Grünes Gewölbe) in Dresden Residence Castle.

In the Sauerland region, Upper Devonian nodular limestones were quarried near Warstein (1860s until 1970s, *Kattenfels*), Brilon (approximately between 1850 and 1914, *Poppenberg*) and Attendorn (17th until the first half of 20th century, *Mecklinghausen Rot* or *Helden Rot*). Westphalian Marbles were mainly of regional importance, although one outstanding example can be found in Thuringia (Altenburg castle, *Helden Rot*).

Lower Carboniferous platform carbonates

During the Lower Carboniferous, an extensive carbonate platform developed along the southern margin of Laurussia (Kohlenkalk facies). Characteristic shallow shelf sediments are dark, bituminous limestones with frequent segments of crinoid columns. Together with Devonian limestones (see previously mentioned), they were used in the Aachen

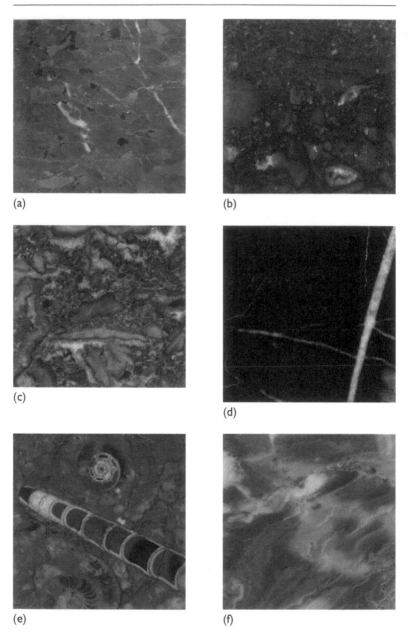

(a)

(b)

(c)

(d)

(e)

(f)

Figure 2.9.2 Important colored "marbles" from Germany. a: Silurian *Goldfleck*; b: Lahn Marble *Unica A*; c: Lahn Marble *Wirbelau*; d: Lahn Marble *Schupbach Schwarz*; e: Saalburg Marble *Edelgrau*; f: Frankenwald Marble *Deutsch Rot*. Each polished slab 10 cm width.

area from Roman times until the end of the last millennium (*Aachener Blaustein*). A little further to the east, Kohlenkalk facies limestone was occasionally quarried in Ratingen (Bergisches Land) near Düsseldorf (19th century). Similar black limestones are also known from the Saxothuringian Zone in the Bohemian Massif. They were quarried in the Royal black "marble" quarry near Wildenfels in the Erzgebirge between the 16th and the 19th centuries (Beierlein 1963) and used, for example, in monuments and buildings in Vienna (Austria) and in Dresden. Around the city Hof (Frankenwald) several quarries existed, including a famous quarry near Schwarzenbach am Wald. The material was processed to several kinds of art objects. Near Döbra and Poppengrün, *Döbraer Schwarz* and *Döbraer Schwarzweiß* were produced in the 20th century.

2.10   Muschelkalk Limestone

The most common Limestone in Germany for monuments and buildings

Roman Koch, Lutz Katzschmann, Klaus Poschlod and Friedrich Häfner

The distinctive grey- to brown-grey-colored buildings and monuments occur throughout Germany from South to North because of the high availability of Muschelkalk Limestone.

The carbonate sediments of the Middle Triassic Muschelkalk were deposited in the whole German Basin. The basin was surrounded by large land areas and it was open to the northern and southern seas. The *Ostkarpatenpforte* and the more important *Burgundische Pforte* allowed the influence of Tethyan faunal elements (Figure 2.10.1).

Sea level fluctuations occurred basin-wide resulting in vertical and lateral facies differentiations during the Muschelkalk period. Recently, this has been documented by sequence stratigraphic studies, which allow the correlation of specific marker beds over large distances.

Because of regional changing environmental conditions in the shallow sea, different facies developments occurred, which are also reflected in varying quality of Muschelkalk building stones.

Geological age

Triassic – Middle Triassic – Tethyan (German Muschelkalk); 240–232 million years.

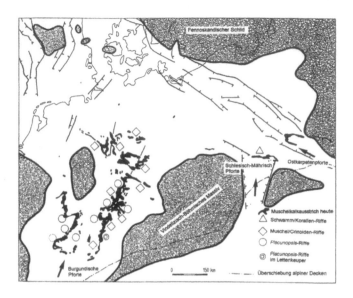

Figure 2.10.1 Paleogeography of the Muschelkalk and the occurrence of so-called "reefs".

Source: Hagdorn et al. 1999.

The Jena Formation is the predominant Formation of the Lower Muschelkalk. It is further subdivided in Lower Wellenkalk Member, Ooid Zone, Middle Wellenkalk Member, *Terebratula* beds, Upper Wellenkalk Member and so-called Schaumkalk beds (oolitic) in the uppermost part.

Within a five-meter thick horizon, two marker beds (*Terebratula* beds) can be traced over the largest distance (up to Poland) of all marker beds in the Jena Formation. The bioclastic limestones of the *Speriferina* beds rich in Crinoids, occurring in the upper third of the Jena Formation, are also easily traceable.

The horizon of the Schaumkalk beds (Figure 2.10.2) shows cross-bedded oolitic limestones commonly used for building stones. The diagenetic dissolution of ooids combined with recrystallization of primary cements results in very porous stone of high quality with a foamy appearance. In Brandenburg, the upper part of the Jena Formation is represented by the Rüdersdorf Formation (predominantly well-bedded oolitic limestones) formed on a regional high caused by salt diapirism.

In the southern and southwestern coastlines of the basin, sandy (Udelfang Formation) and local dolomitic intercalations occur (Freudenstadt Formation, Eschenbach Formation).

(a)

(b)

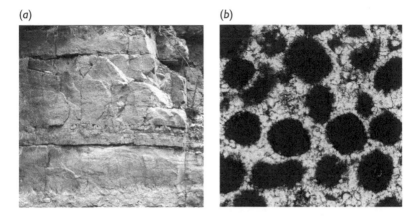

Figure 2.10.2 (a) Schaumkalk beds of up to 30 cm thickness showing low angle cross-bedding. (b) Schaumkalk with dissolved ooids and recrystallized marine cements (size of ooids < 1 mm; thin section microphotograph, // Nic.).

(a)

(b)

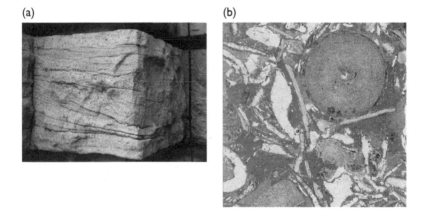

Figure 2.10.3 (a) Muschelkalk block with marked disconformities caused by alternating small sedimentary units. Height 40 cm; Nuremberg main railway station. (b) Crailsheim Muschelkalk from the Quaderkalk Formation with cross section of crinoids (modified from Grimm 2018; thin section microphotograph // Nic, size 8 mm).

Only the relatively homogeneous micritic dolomites and dolomitic limestones of the Karlstadt and Diemel Formation (Middle Muschelkalk) were used to a small extent as building stone.

(a)   (b)

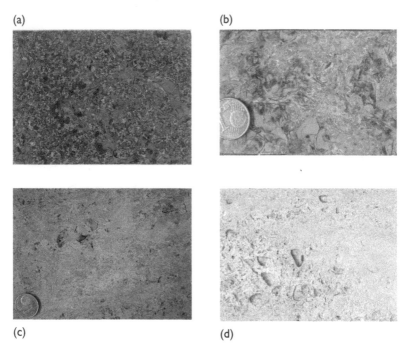

(c)   (d)

Figure 2.10.4 (a) *Sellenberger Muschelkalk* (10 × 8 cm); (b) *Kirchheimer Goldbank*
           (both Bavaria); (c) *Oberdorlaer Muschelkalk* (Thuringia); (d) *Elm-
           Kalkstein* (18 × 13 cm, Lower Saxony).

Source: Photos by BGR.

The Upper Muschelkalk consists of the *Trochitenkalk*, Meissner (incl.
Nodosus beds) and Warburg Formation. It is most important for the
extraction of building stones.

The uppermost Rottweil Formation in SW Germany characterizes a
coastal, dolomitic sedimentation (*Trigonodus* Dolomite).

The Trochitenkalk Formation (thickness up to 40 m) shows an alternation of bioclastic, oolitic and micritic limestones with common crinoids
(*Encrinus liliiformis*; fragments of sea lilies).

The shallow water deposits of the so-called Quaderkalk Formation,
which is a lateral time-equivalent of the Meissner Formation, consist of
beds of meters' thickness. Within an area of about 60 km length and 30
km width striking in NNW-direction, it reaches a thickness of up to 20 m.

They predominantly consist of mollusk fragments and ooids. Large-
scale cross bedding of decimeters' to meters' thickness is characteristic.

So-called "reefs", occurring in this area (Figure 2.10.1) are no reefs like recent coral reefs. Algae and hydrozoans formed Muschelkalk "reefs" settling within and stabilizing high energy carbonate sand piles.

Occurrence, quarrying and characteristics

Quarries occur in the Jena Formation, but most excellent building stones are extracted from the *Terebratula* beds in Thuringia and from the *Terebratula* beds and Quaderkalk Formation in Bavaria and Baden-Württemberg.

The Schaumkalk beds are commonly quarried in Oberdorla (Thuringia), occasionally in Freyburg (Saxony-Anhalt), and near Bad Neustadt a. d. Saale (Bavaria); in former times also near Meiningen (Thuringia) and Rüdersdorf (near Berlin).

The *Oberdorlaer Schaumkalk* consists predominantly of bivalve debris. It shows reddish-brownish, porous areas (up to 10%), high water uptake and is very resistant against weathering. Most prominent examples for its use are the police headquarters in Berlin-Charlottenburg, the monument in the concentration camp (KZ) in Buchenwald near Weimar and the façade of the new library of the Augustinian Monastery in Erfurt. The *Meininger Schaumkalk* was also used for large buildings and monuments (Elisabethenburg, Brahms Monument).

The *Freyburger Schaumkalk* and the *Rüdersdorfer Muschelkalk* represent *"Schaumkalk"* in high energy shallow marine oolitic environments, whereas other limestones named Schaumkalk have high matrix content and fossil debris. The Naumburg Cathedral (see Chapter 1.7) is the most prominent example for *Freyburger Schaumkalk*.

Beds of the Lower Muschelkalk were also used in other locations for important buildings, as the Jena Muschelkalk (*Terebratula* beds, University in Jena) or limestones from Lower Saxony, for example, *Elm-Kalkstein* (abbey church in Königslutter, Behrens buildings at Alexanderplatz in Berlin) and *Osnabrücker Wellenkalk* (Waldbühne Berlin).

Most important building stones of excellent quality are found in the Upper Muschelkalk in Bavaria and Baden-Württemberg.

Especially the thick-bedded Quaderkalk facies is often used inside and outside Germany. It is quarried southward of Würzburg in an area between Kirchheim, Krensheim, Crailsheim and Ochsenfurt showing facies variations in limestones rich in fossil debris and limestones rich in micritic matrix (*Kirchheimer Kernstein, Kirchheimer Blaubank, Kirchheimer Goldbank*). The characteristic color (grey, grey-bluish, ocher) are caused by different Fe-minerals in the micritic matrix.

Outstanding buildings and monuments were made of these limestones as there are the Olympic Stadium and the Pergamon Museum in Berlin (see Chapter 1.2), the railway main stations in Stuttgart and Nuremberg, and university buildings in Munich and Würzburg.

Other limestones from the Upper Muschelkalk used for building stones were the *Trochiten-Kalkstein* from Polle and Erkerode (both Lower Saxony), and Lippe Mountains (North Rhine-Westphalia) and, especially in the Roman Era, in the southern Eifel Mts. near Trier (see Chapter 1.8).

Technical parameters

Muschelkalk limestone is characterized by two main parameters, which are responsible for the good weathering stability. On the one hand, the general intensive cementation results in high to very high compressive and flexural strengths. A certain amount of micritic matrix alters this parameter only moderately. On the other hand, the commonly very high open porosity causes the very good stability against freezing-taw cycles. For problems of chemical weathering, see Chapter 1.7.

Figure 2.10.5 Lion sculpture made of Terebratel limestone, main building of the Friedrich Schiller University Jena.

Figure 2.10.6 Augsburg Synagogue (1917), socle made of *Kirchheimer Muschelkalk Kernstein.*

Figure 2.10.7 Olympic stadium in Berlin (1936) – all façades clad with Bavarian Muschelkalk.

Source: Photo by U. Schönitz.

Table 2.10.1 Technical Parameters from Grimm (2018) and from Recent Authors

	Open Porosity [vol.%]	Water Absorption (atm.) [wt.%]	Compressive Strength [N/mm²]	Flexural Strength [N/mm²]
Oberdorla	19.6–20.4	2.20–4.13	24–57	6.2–6.7
Jena	7.1–18.9	1.30–3.70	30–57	3.7–9.1
Meiningen	1.2–9.1	0.64–3.04	26–59	7.4
Freyburg	17.6–38.2	2.75–5.70	18–51	–
Rüdersdorf	13.47	2.69	–	–
Osnabrück	4.54	1.38	–	–
Elm-Kalkstein	24.6–31.8	4.1–4.07	21–90	2.6–5.8
Kirchheim Kernstein	2.5–3.5	0.60–1.70	40–50	8–10
Kircheim Goldbank	0.88–1.85	0.27–0.51	–	–
Kirchheim Blaubank	0.90–1.80	0.27–0.50	113–151	8–12.9
Crailsheim	3.5–11.2	1.82–2.46	–	–
Krensheim	13.8–13.8	2.40–2.44	65	10.2

2.11   Jurassic limestones

Solnhofen Limestone (Solnhofener Plattenkalk) and Treuchtlingen Limestone (Treuchtlinger Kalkstein)

Roman Koch and Anette Ritter-Höll

Brief geological description

Treuchtlingen Limestone ("Treuchtlingen Marble") and Solnhofen Limestone (Plattenkalk; lithographic limestone) are famous stones for buildings, monuments and bas-reliefs for hundreds and thousands of years. Both can be used inside and outside depending on the quality of the specific beds.

The small river Altmühl deeply cuts into the sequence of Upper Jurassic limestones of the southern *Fränkische Alb*. Because of the stratigraphic position and the morphology of the area, the Solnhofen Limestone occurs on the top of the landscape. The Treuchtlingen Limestone is predominantly quarried along the steep slopes of the Altmühl Valley and associated smaller valleys. The Upper Jurassic beds are nearly horizontal, and single beds of the Treuchtlingen Limestone can be traced over dozens of kilometers (Beyer & Grimm 1997) whereas the vertical sequence shows a characteristic alternation of varying limestone types (facies types) bed by bed.

Geological age

Mesozoic – Upper Jurassic (Malm) – Solnhofen Limestone: Tithonian (Malm Zeta 2–3) Altmühltal Formation (152–145 million years).

Treuchtlingen Limestone (Treuchtlingen Formation): Kimmeridgian (154.5–153.5 million years).

Solnhofen Limestone

The Solnhofen Limestone, composed of limestone beds of centimeter to about three decimeters thickness, is one of the most famous limestone of Southern Germany, not only because of the occurrence of the *Archaeopteryx*, which was found in the area of Eichstätt. The "blue" Solnhofen Limestone from the Mörnsheim quarry is the original stone used for lithography (lithographic limestone) invented by Alois Senefelder in 1796 for printing onto paper. The Solnhofen Limestone can be used because of its extraordinary facies and petrophysical parameters (Keupp 1977; Koch 2007). In the most prospering period, 28 quarries were active, predominantly in the "blue" Solnhofen Limestone. Recently, the yellowish Solnhofen Limestone occurs nearly exclusively; it shows common dendrites and rare fossil relics (ophiura, fishes) on bedding planes.

The Solnhofen Limestone was used for floors in the interior of numerous churches and historical buildings as well as for the construction of platy roof systems, which are characteristic for traditional houses in the southern *Fränkische Alb* area. Because of its extremely fine crystal size, it can be sculptured very precisely and was used for thousands of bas-reliefs (epitaphs).

Figure 2.11.1 Solnhofen Limestone together with Ruhpolding Limestone in the Munich Residence.

Occurrence, quarrying and use

The Solnhofen Limestone occurs in a very limited area of 30 × 70 km in the southern *Fränkische Alb* between Solnhofen in the West and Kelheim in the East.

The most characteristic type of Solnhofen Limestone is found in Malm Zeta 2 at the localities of Langenaltheim, Solnhofen, Mörnsheim, Eichstätt, Daiting and Zandt. The limestone was deposited in local basins (so called "Wannen"), which were separated from each other by areas of massive limestone ("reef deposits").

Historical quarrying was by hand, hoe and shovel. The limestone plates were transported by wheelbarrow, later using a small quarry-railway. Worker groups of three to six persons separated the different layers carefully bed by bed and sorted it according to the sound of the hammer (bright = good; dull = bad). Subsequently, the format of the plates was marked using a normalized wooden frame and a special hammer for taking off the protruding parts of the plates.

Today, the Solnhofen Limestone is quarried in the area of Langenaltheim, Mörnsheim, Solnhofen, Eichstätt and Zandt. In Mörnsheim, the Solnhofen high-quality beds (*Flinze*) show the greatest thickness with 32 cm. In other locations (e.g., Eichstätt and Zandt), only very thin beds of *Flinze* occur (Skinny *Flinze* 10–20 mm, Leaf *Flinze*: 3–9 mm, Paper *Flinze*: 1–2 mm).

The total thickness that can be quarried is 50–60 m in the Mörnsheim area whereas much lower thickness occurs in the Eichstätt (20–25 m) and Daiting area (5–10 m).

Figure 2.11.2 Manual quarrying of Solnhofen plates.

In the Lichtenberg quarry of the Mörnsheim area ("*Zugspitz* quarry"), 290 beds were counted consisting of 163 high-quality *Flinze* (thickness 5 mm to 32 cm) and 127 *Fäulen*. The quarrying delivers 90% plates of small dimensions. Large dimensions (40 × 40 cm; 60 × 60 cm) are still used for sacral buildings, stairs and covering; consequently, the price depends on the size of plates. Today, the surface characteristics of Solnhofen plates are different depending on the quarry. In the Eichstätt area, the plate surfaces are extremely fine (homogenous) whereas so-called tiny peaks occur in the Solnhofen *Maxberg* quarry. The term "quarry rough" refers to this difference. "Partly honed" refers to the moderate, and "fine-honed" to the complete off-grinding of this peaks.

Petrographical characteristics

The Solnhofen Limestone is characterized by a close alternation of tight, micritic beds (*Flinze*) from a few centimeters to decimeter thick, and thin intercalations of clay-rich layers (*Fäulen*; Figure 2.11.3).

The micrite reveals slightly different particle (crystal) sizes in *Flinze* and *Fäulen*, which is interpreted to be primary without any diagenetic alteration (Keupp 1977). A certain amount of organic matter in the Solnhofen Limestone is reflected by the original grey to greyish-blue color. The change of this color to yellowish and white is caused by oxidation of this fine dispersed organic matter.

The Solnhofen Limestone shows a predominance of nektonic and planktic organisms and traces of fine benthonic fossils. The origin of the micrite was explained by chemical precipitation, bacterial-induced precipitation and disintegration of coccolithophorids. Two micritic facies types of Solnhofen Limestone are predominantly used for floors and bas-reliefs.

The homogenous, extremely tight, cream-colored to yellowish Solnhofen Limestone is the most preferred stone (Figure 2.11.4(A) and (B)). Thin fractures, healed by fine granular calcite, can occur and do not alter the good stone properties. The second type shows whitish spots, irregularly distributed in the micrite (Figure 2.11.4(C)) or oriented along small fractures. These spots are microporous areas (Figure 2.11.4(D), blue-stained) in which the micrite is recrystallized to slightly larger crystals (microsparite). Reaction seams were formed due to moisture kept in these spots for longer time.

Weathering resistance

Epitaphs and very delicate bas-reliefs of Solnhofen Limestone are very resistant against weathering when treated with specific agents.

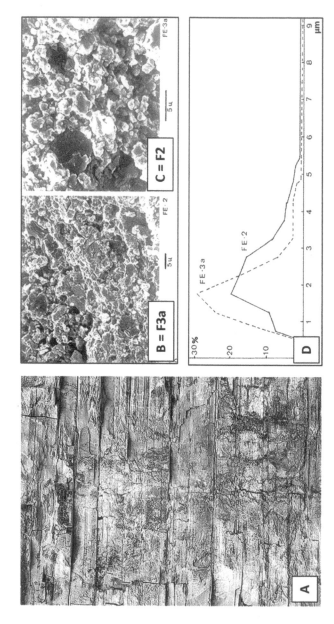

Figure 2.11.3 (A) Characteristic view of the Solnhofen Limestone from the Mörnsheim quarry with micritic limestone beds (*Flinze*) of 2 to 20 cm thickness and thin marly intercalations (*Fäulen*). Distribution of particle size (D) in *Flinze* (B) and *Fäulen* (C) in Solnhofen Limestone from the Lichtenberg quarry.

Source: Keupp 1977.

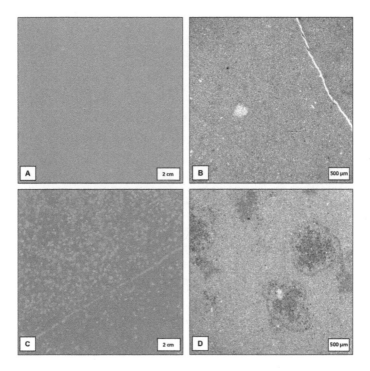

Figure 2.11.4 Solnhofen type 1 (A – photograph of a plate; B microphotograph, parallel light) and type 2 (C and D). Type 1 is an extremely
homoge nous micrite (B). Type 2 contains whitish spots that are
small porous (blue stained) areas with local reaction seams due to
moisture kept therein (D).

Damages can occur by fracturing, breaking out of sharply limited parts,
by splitting off fine-bedded parts, and by surface decay in very fine particles of silt-size. This points to varying types of the Solnhofen Limestone
of different quality.

Treuchtlingen Limestone (Treuchtlingen "Marble")

The Treuchtlingen Limestone ("Jura Marble") consists of well-developed
beds of 25 cm to 1.40 m thickness forming a quarried sequence of about
28 m total thickness. In a standard profile of the Treuchtlingen area, it consists of at least 27 beds (Figure 2.11.5). The cream-colored to light brown

Figure 2.11.5 Treuchtlingen Limestone with 27 beds here in this quarry, a part of which can be used for building stones outside and a part for constructions inside (height about 25 m; Petersbuch quarry).

and grey limestone beds of high density reveals varying amounts of different microfossils changing from bed to bed. The Treuchtlingen Limestone is overlain by dolomitic beds, which are also used for buildings stones.

Occurrence, quarrying and use

The Treuchtlingen Limestone was first quarried in the area of Treuchtlingen and Pappenheim. Limestones, which can be polished, were called "marble" in the stone industry. Thus, the term *Treuchtlinger Marmor* (*Juramarmor*) was used.

Later, quarries in the region of Titting, Pollenfeld, and Eichstätt (Obereichstätt; Blumenberg) were opened and the term was changed to Treuchtlingen Limestone. Today it is quarried in about 30 quarries, which are not continuously active.

About one-third of the 30 beds can be used for inside floors (beds No. 1–6; thickness 65–140 cm) and one-third for monuments under atmospheric conditions (beds No. 8–30 with some unusable beds; thickness 65–130 cm; frost resistant). In general, colors that are more greyish occur in the lower part and lighter brownish colors in the upper part of the sequence, probably caused by deep reaching surface weathering. Beds of yellowish or golden and whitish color are recently preferred for facades worldwide. Grey colors are predominantly used for floors inside.

Additionally, different surface treatments of plates (sawed, polished, sandblasted, bushhammered, honed a. o.) are applied to produce rocks with a large variety of optical appearance of different surfaces. Furthermore, sawing the blocks parallel or vertical to bedding planes results in different characteristics. Therefore, the name *"Travertin"* was given to whitish to light cream-colored plates sawed vertical to the bedding plates, resulting in sedimentary structures resembling original travertine.

The limestone has been used in very many public and private buildings throughout Germany for many decades as flooring, staircases and as windowsills. It is also used for façades of skyscrapers in China, U.S.A. and Emirates as well as in Berlin, London and Munich.

Colors available are labelled *Jura Gelb* (yellow), *Jura geblümt* (flowered), *Jura Grau* (grey) and *Jura Gemischt* (mixed).

The limestone is also used as a building and sculptural stone within and around the mining region.

Petrographical characteristics

The Treuchtlingen Limestone was deposited in a shallow marine carbonate platform environment and is predominantly composed of two facies types (Kott 1989).

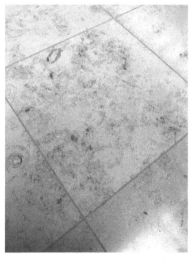

Figure 2.11.6 Portal made with Jurassic Limestone in Munich.

Figure 2.11.7 Jura Grey at the floor of the new Berlin Airport

Source: Photo by C. Barthelme.

A matrix-rich (micrite) bioclastic wackestone with intraclasts of mm – cm size also shows ammonites deposited parallel to bedding planes. It reflects deposition under moderate water energy.

A bioclastic facies rich in peloids and microbial crust contains abundant peloids, small ooids, oncoids and commonly so-called Tubiphytes, which are the microfossils that are most characteristic for the Treuchtlingen Limestone (Figure 2.11.8). Furthermore, benthic foraminifera, serpulids, bryozoans and fragments of molluscs and echinoids occur.

Some beds show primary pores healed by granular calcite. A certain amount of open porosity occurs, influencing the capillary water uptake and the stability against freeze/thaw cycles.

Treuchtlingen Limestone used for building stones commonly reveals an alternation of more micritic, fine layers and coarser layers. Coarse layers are composed of peloids, lithoclasts and so-called "white flames" (arrows), which are the microorganism "*Tubiphytes*" (Figure 2.11.8(B)). The grey Treuchtlingen Limestone is denser and commonly contains fragments of sponges floating in micrite, containing thin fragments of shells ("Filaments").

Weathering resistance

The weathering resistance of Treuchtlingen Limestone depends on the sedimentological and diagenetic parameters of different beds (matrix content, porosity, packing of components, stylolites, dolomite, dedolo-mit, silification; Ritter-Höll 2005) as described earlier.

Technical parameters

The technical specifications of Solnhofen and Treuchtlingen Limestones are as follows.

Table 2.11.1 Technical Parameters of Solnhofen and Treuchtlingen Limestones

	Solnhofen	Treuchtlingen
Open porosity [g/cm^3]	2.60–3.70 (4.77*)	3.38
Water absorption (atm.) [wt.%]	1.1 (1.47*)	1.10
Compressive strength [MPa]	180–240	100–140
Flexural strength [MPa]	5–15	4–18

Source: Data from Solnhofen industry; * data from Grimm 2018, modified.

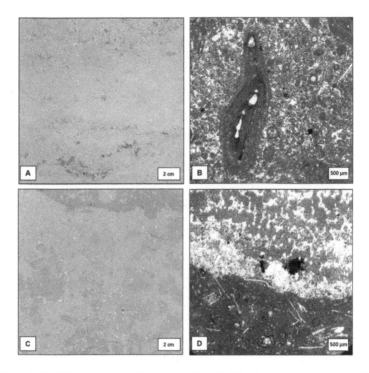

Figure 2.11.8 Treuchtlingen Limestone. Type I (A) plate, primary pores (dark grey) are closed by granular calcite. (B) Longitudinal section of "Tubiphytes" with central foraminiferal chambers (white) floating in a peloid packstone (microphotograph, parallel light) The Grey Treuchtlingen Limestone (C; type 2) contains fragments of sponges and of thin shells (Filaments; D).

2.12   Roof slates

H. Wolfgang Wagner

Brief geological description

The German roofing slates are very low-grade metamorphic silt and clay schists with a transverse schistosity that can be split up to 5 mm. In the past, colored slates (green and purple) were used for roofing purposes, but today only black slates are still quarried. Fine-grained sediments were deposited in a marine back-arc basin along the southern passive margin of the paleo continent Laurussia. Roofing slate and the slaty cleavage were formed during the Hercynian Orogeny (about 320 to 280 Ma.). The most important deposits are located in the Rheno-Hercynian Zone of the Central European Variscides; other deposits also existed in the Saxothuringian Zone. Today two slate types are extracted as roof slate: Middle Devonian (dm) slate: black with > 5% Carbonate, and Lower Devonian (du) slate: black with < 2% Carbonate.

Both are presented in detail.

Geological age

Palaeozoic – Devonian

Middle Devonian – Eifelian – Fredeburg Formation – Wissenbach-Schiefer; 390 Ma.

Lower Devonian – Emsian – Altlay Formation; 408 Ma.

Occurrence, quarrying and use

The Rheno-Hercynian is part of the Variscides and extends from the Harz Mountains to northeast to the Rhenish Slate Mountains (*Rheinisches Schiefergebirge*) and to the Ardennes in the southwest, with several roof slate deposits. In this area, slate roofs are characteristic for the man-made landscape.

After closing of the Rheic Ocean in the Lochkovian period (420 to 414 Ma.), large-scale shear and transformation processes resulted in the

Figure 2.12.1 Slate ornament in the German roofing style at a façade.

stretching and opening of a narrow Rheno-Hercynian Ocean (408 to 405 Ma.) with the first oceanic crust or the formation of MOR (= Mid-oceanic ridge type) basalts.

The formation of the symmetrical Rheno-Hercynian Rift Basin located in the northwest started in the Lochkovian and Pragian stages (414 to 408 Ma.). The entire basin was divided into horsts and troughs, with the later roofing slates being deposited in the respective trough areas. The process of subduction ended with the continent collision of the Mid-German Crystalline High and the Rheno-Hercynian. The collision was associated with a deformation migrating from the southeast to the northwest with predominantly NW-vergent folds and thrusts as well as the uplifting of mountains at the end of Namurian (313 Ma.). The folding and the associated slaty cleavage and very low-grade metamorphosis of this orogenesis gave the pelites their suitability for roof slate (Figure 2.12.2).

The Lower Devonian roof slates are located in a zone extending from the Taunus in the north-east, the Middle Rhine to the Eifel (*Moselschiefer*, Eastern Eifel with the Katzenberg and Margareta mines, closed in 2019) and the Hunsrück (Altlay mine still active, newly developed Frühberg 3 opencast mine near Bundenbach) to the southern Ardennes in the southwest. During the production process, the world-famous, well-preserved fossils from Bundenbach were discovered while splitting the slate.

The Middle Devonian roof slates occur in the Moselle Syncline near Trier in the southwest, turn into the Lahn-Dill Syncline (deposits of Wissenbach and Rupbach) and end in their north-east extension in the Upper Harz Devonian anticline near Goslar. On the right bank (east) of the river Rhine, north of the Siegen thrust, there is an area with Middle Devonian slates (still active Magog mine near Fredeburg). In addition to the Devonian slate, there are or were other deposits in the Lower Carboniferous of the Saxothuringian near Geroldsgrün (Lotharheil)/Bavaria and Lehesten and Unterloquitz/Thuringia.

In Germany, the mining and extraction of roof slate takes place in underground mines in Altlay and Magog. The slate is extracted using specially developed mobile sawing and splitting machines. The raw blocks are transported using LHD (Load-Haul-Dump) technology (*Gleislostechnik*) (Figure 2.12.3).

Further processing follows by sawing the raw blocks, splitting to a thickness of 5 mm and trimming of the slate. The free-hand dressing and trimming, especially of *Altdeutsch* (Old German) slate slabs, has a long tradition and requires special craftsmanship (Figure 2.12.4). The tradition of free-hand dressing of slate is now in danger. There are plans by some slate suppliers to produce such formats abroad with NC-controlled machines.

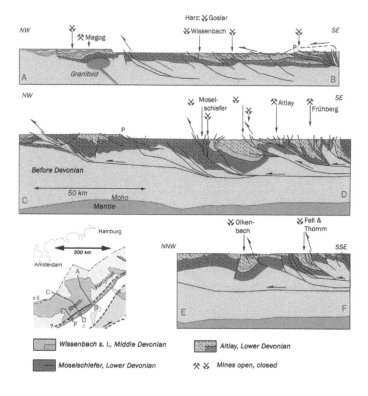

Figure 2.12.2 Geological profiles of the Rheno-Hercynian with the most important slate mines.

Petrographical characteristics

Roof slate mainly consists of phyllosilicates (muscovite/illite, paragonite/brammallite and chlorites), as well as quartz. Other accessory constituents such as carbonate, non-carbonate carbon (= nc carbon) and ore minerals can influence the properties and color of the slate. The mineralogical composition of the roof slates mined today is as follows: mica, 32–42%; chlorite, 14–20%; illite, 6–8%; quartz, 27–31%; carbonate, 1–10%; and accessory feldspar, pyrite and other ore minerals.

The most important characteristic of a slate is its cleavage, which originates from a very fine, petrographic parallel structure. This structure is examined microscopic in a thin section. The number of mica (= phyllosilicate) layers per millimeter (a) and the average thickness of the mica

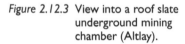

Figure 2.12.3 View into a roof slate underground mining chamber (Altlay).

Figure 2.12.4 The traditional free-hand dressing of slate, especially of variable *Altdeutsch* (Old German) (Altlay).

layers (b) are examined (Table 2.12.1). The quantity value is calculated from both values: (a) × (b) × 10.

Technical parameters

Table 2.12.1 shows the diversity of structural and technical parameters. All slates meet the requirements of the European roofing slate standard EN 12326–1. They are also considered frost-resistant due to water absorption < 0.6%. In Magog slates, there is only a slight difference between the longitudinal and transverse flexural strength; in Altlay and *Moselschiefer*, they are almost the same.

These therefore do not have a "grain" (French: longrain, German: Faden) that would make the production of round formats (such as "Altdeutsch") more difficult.

In the Rhine and Moselle regions (e.g., Xanten, Mayen, Koblenz), slate roofs exist since the Roman era and were as common as the Roman tiles (Figure 2.12.5).

The first written record of these slate roofs is dated from 1150 to 1158 by Hildegard von Bingen. The slater craft was of great importance in the Middle Ages, when slater guilds existed in cities with UNESCO World Heritage Sites, such as in Trier (around 1100), Goslar (around 1300) or Cologne (1397). This guild tradition still exists today in the form of the "*Vereinigung der rechtschaffenen fremden Zimmerer- und Schieferdeckergesellen Deutschlands*" (Association of Righteous Foreign Carpenters and Slaters in Germany) with about 150 traveling workers. Later on, fire regulations in cities led to the prohibition of all soft, that is, flammable roofing materials such as straw, reeds and wooden shingles. They

Table 2.12.1 Structural and Technical Parameters of Roof Slates

	Mica Layers/ mm	Thick- ness of Mica Layers μm	Quantity Value	Flexural Strength [MPa] longitudinal	Flexural Strength [MPa] transversal	Water Absorption (atm) [wt.%]	Raw Density [g/cm³]
Altlay	106	4–9	3.5–9.4	50	56	0.28	2.80
Magog	72–86	5	2.7–5.8	50	38	0.50	2.77
Mosel-schiefer	88–98	3–5	2.8–4.7	54	52	0.22	2.78

Source: After Wagner 2020.

regulated by law hard roofing materials and therefore slate. Slate use peaked in the late 19th century. At that time, hundreds of roofing slate mines were operating in Germany. In the low mountain range, especially near the mines, slate roofs and façades are still common to this day. Outside of these classic slate areas, slate was only used in cities to cover particularly complex roof shapes or in case of existing trade connections to slate areas (e.g., port cities). Today, roof slate of the two active mines Altlay and Magog is available for important heritage buildings. Alternatively, roofing slates from NW Spain are mostly used.

Weathering resistance

In Germany, both the roof slate from the Lower Devonian and the roof slate from the Middle Devonian have sufficient weathering resistance. Due to the different carbonate contents, the Lower Devonian slate is usually color-stable (= unfading), while the Middle Devonian slate is usually not (= fading).

UNESCO World Heritage Sites

The Roman Upper German-Rhaetian Limes stretches over 550 kilometers from the Rhine to the Danube. Many of its former watchtowers and associated forts were evidently covered with slate and were also reconstructed with it (Figure 2.12.5).

Slate roofs on half-timbered houses and castles characterize the Upper Middle Rhine Valley. Slate roofs are also very common on historic buildings in the old towns of Bamberg (Figure 2.12.7), Goslar, Classical Weimar, Regensburg, Quedlinburg and on the Luther Memorials in Eisleben and Wittenberg. The castle and church buildings in Potsdam, Würzburg Residence, Augustusburg and Falkenlust at Brühl, the Cathedral and

Church of Our Lady in Trier, the monasteries Lorsch and Maulbronn (ridge turret) and the Lions Castle in the Bergpark Wilhelmshöhe have slate roofs as well. The majority was traditionally covered in German roofing style, for example, *Altdeutsch* and not with rectangular slates (Figure 2.12.1 and 2.12.7). There are exceptions, however, such as the Aachen Cathedral, with its entrance hall covered in *Altdeutsch*, but the choir hall covered with rectangles that are common in neighboring Belgium. Both roof coverings also appeared earlier in port cities: in addition to the German slates transported over rivers, rectangular slates imported from Wales/U.K. occurred as well. Examples include the Speicherstadt in Hamburg and roofs in Lübeck (Figure 2.12.6). The Welsh Slate is now recognized as a Global Heritage Stone Resource.

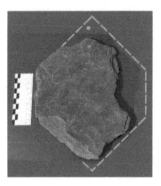

Figure 2.12.5 Roman hexagon slate, partially reconstructed (villa rustica Buchen near Limes).

Figure 2.12.6 English rectangular slate from Wales in Lübeck near the Holstentor (approx. 100 years old).

Figure 2.12.7 Michaelsberg Monastery newly roofed with slate from Altlay, Bamberg Old Town (photo by Theis-Böger).

2.13   Rhenish Basalt Lava (*Rheinische Basaltlava*)

Friedrich Häfner

Brief geological description

Intensive volcanic activities took place in the Eifel Mts. in the younger geological history, during Cenozoic. Overall, about 370 volcanic structures are proved (LGB 2005). The volcanic rocks are classified as nephelinites, leucitites, basanites and tephrites as well as phonolithes and tuffs. In terms of marketing, the basaltic rocks are, independent of their precise composition, named, for example, *Mayener Basaltlava*, *Mendiger Basaltlava* and *Plaidter Basaltlava*. The collective name for all these volcanic rocks is Rhenish Basalt Lava.

Geological age

The volcanoes of the Eifel Mts. are evidence of the youngest volcanic area in Europe. The activities began 30 million years ago (Tertiary) and ended in the Quaternary period. The basalt lavas in the eastern Eifel Mts. are less than 600,000 years old. The last eruption took place in the Laacher See area around 12,900 years (LGB 2005).

Occurrence, quarrying and use

The lava streams had two central areas: one in the eastern and one in the western part of the Eifel Mountains. Most of the lava streams contain big pillars, some of them with diameters up to 3 m. The dimension stones are extracted by blasting. Currently, five companies exploit the rocks in quarries, which are not operated all the year round. Main locations are Mendig and Mayen in the eastern Eifel Mts. (Figure 2.13.2). Schumacher and Müller (2011) calculated the area of deep mining in Mendig to an extent of 3 km² and a volume of extracted material up to 24 Mio. m³. The center of the Mayen quarrying area, the so-called "*Mayener Grubenfeld*", has similar dimensions. The Romans already extracted the basalt lava in Mayen, Andernach and Plaidt. The total amount of basalt lava exploited by the Romans is roughly estimated at three Mio. m³ (Mangartz 2000).

Most of the industrial extracted basalt lava is indeed used as aggregates in the asphalt and concrete industry as well as in hydraulic construction. The Rhenish Basalt Lava is very suitable for producing all kinds of ornamental stones and building stones. Nowadays,

Figure 2.13.1 Relics of Roman millstone production in the Mayen Mining Area.

Figure 2.13.2 Two flows of Rhenish Basalt Lava with an interlayer of loess exposed in the quarry "*Stürmerich*" in Mendig (Eifel) (2010).

Figure 2.13.3 Tower erected with Rhenish Basalt Lava (medieval surrounding walls of Mayen, Eifel), 14th century.

the dimension stones are used for restoring historic monuments, solid buildings, for sculptures, facing tiles, pavement, flagged floors, fountains, horticulture and landscape construction. Very important examples for their use in the past are millstones, which were produced since Roman times and were exported to all parts of the Roman Empire

(Figure 2.13.1). Outstanding applications are the Roman Bridge in Trier (see Chapter 1.8, Figure 1.8.8), the Genoveva Castle in Mayen, the portal of the city hall of Mayen and the medieval surrounding walls of Mayen (Figure 2.13.3).

Examples of younger buildings are the Gertraudenbrücke in Berlin, an office building in Leipzig (façade), the thermal baths in Längenfeld and the Museum Moderne Kunst (MUMOK) in Vienna (both Austria).

Petrographical characteristics

At this point, the phonotephrites and tephritic phonolithes of the Mayen Basalt Lava (*Mayener Basaltlava*) are presented as representative examples (Häfner 2008).

At first sight, the basalt lava seems to be relatively homogeneous (Figure 2.13.4). In detail, differences appear regarding colors from blue to greyish blue and to brownish grey; the pores vary in size, shape, quantity and distribution. Locally, inclusions of different rocks like crystalline schist, sandstone, greywacke as well as limestones and marlstones can be included. The pores are often filled with different minerals like nepheline or leucite (Häfner 2008).

Microscopically, the basalt lava is characterized by a holocrystalline porphyric structure: inclusions of pyroxene (augite, partly green), intensively altered mafic minerals (hornblende, biotite) and sporadic olivine are "swimming" in a fine-crystalline homogeneous matrix, a mixture of pyroxene (augite), feldspar (plagioclase, K-feldspar), foides (nepheline, leucite) and opaque ore minerals. Biotite, apatite and glass portions appear accessory. Locally, corrosive and partly altered crystals of quartz and feldspar occur as inclusions in the rocks.

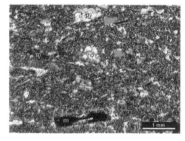

Figure 2.13.4 Specimen of Rhenish Basalt Lava, size 20 × 30 cm.

Figure 2.13.5 Thin section of Rhenish Basalt Lava, sample no. 03137.

Table 2.13.1 Technical Parameters of the Rhenish Basalt Lava from the Mayen Area

Bulk Density [g/cm³]	Compressive Strength [MPa]	Flexural Strength (⊥) [MPa]	Water Absorption (atm) [wt.%]	Porosity (eff) [vol.%]
2.24–2.62	50–92	11.3–13.3	2.4–6.2	11.9–22.3

Source: Häfner 2008; Grimm 2018.

The size of the crystallized minerals in the matrix is varying between 0.006 und 0.15 mm (mostly 0.06 mm). Porphyric inclusions have diameters up to 1 mm (Figure 2.13.5) (Häfner 2008). The mineralogical composition of the Mayen lava stream is as follows (3 samples, Häfner 2008): matrix 67–85%; inclusions (mainly pyroxene, little mafic minerals, less olivine) 7–12%; and pores 3–25%.

Technical parameters

The Rhenish Basalt Lava is resistant to sunburn and during freeze/thaw cycles.

Weathering resistance

In spite of their relevant porosity and the resulting water absorption of partly more than 6 wt.%, the basalt lava can be confirmed as having very good resistance against weathering, including frost. Even at buildings from Roman or medieval times, no severe damages can be observed.

2.14   Drachenfels Trachyte (Drachenfels-Trachyt)

Esther von Plehwe-Leisen and Hans Leisen

Introduction

The trachyte of the *Drachenfels* (DT) was quarried in the *Siebengebirge* near Bonn. It is a light volcanic rock with porphyric texture and large sanidine feldspar crystals.

DT was already quarried in Roman times, but since medieval times, it became the most important building stone in the Rhineland and neighboring areas. The material turned so popular that the extension of its

Figure 2.14.1 View of the Drachenfels in the *Siebengebirge* and the Rhine River.

use followed not only the rivers Rhine, Lahn and Moselle but also chal-
lenging land transport routes. The famous World Heritage Site Cologne
Cathedral is the most prominent example of the use of DT as a building
stone.

Nowadays, DT shows manifold weathering degradations due to its
long lifetime and exposure in polluted areas.

Geological age

Tertiary – Miocene – 25.6 million years.

Geological and petrographical description

During the late Paleogene period, volcanic eruptions in the *Siebengebirge* occurred in connection with the subsidence of the Cologne Embayment as part of the Lower Rhine Graben.

The intracontinental Siebengebirge Volcanic Field (SVF) shows
a great variety of volcanic rocks deriving from at least two parental
magmas. The silica-saturated DT formed by differentiation and crustal
assimilation processes from an alkali-basaltic magma ca. 25 Mio years

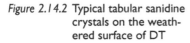

Figure 2.14.2 Typical tabular sanidine crystals on the weathered surface of DT

Figure 2.14.3 Sanidine crystal surrounded by smaller phenocrystals of feldspar and fine crystalline matrix (thin section N+).

ago (Przybyla et al. 2018). The DT magma ascended and got stuck in older volcanic tuff layers and cooled down as an intrusive lava dome (Cloos & Cloos 1927; Berres 1996). The tuff layers weathered and were eroded, thus exposing the trachyte dome as the mountain called *Drachenfels*. The scientific exploration of the *Siebengebirge* and the *Drachenfels* already began in the 18th century (Schwarz 2014: good compilation).

The DT shows a porphyric fluidal texture. Tabular alkali feldspar (sanidine) crystals can reach lengths of up to 7 cm, and they are oriented in the direction of the magma flow (Cloos & Cloos 1927). Besides sanidine, phenocrysts of plagioclase, biotite, and augite are present (Figure 2.14.2 and 2.14.3). The fine-grained matrix mainly consists of feldspar and quartz. Additionally, magnetite, apatite, augite and chlorite can occur (Frechen & Vieten 1970; Simper 2018).

DT shows low water absorption and medium dilatation and good mechanical strength (Table 2.14.1).

Table 2.14.1 Petrophysical Properties of the Drachenfels Trachyte

Compressive Strength [MPa]	Flexural Strength [MPa]	Water Absorption (atm) [wt.%]	Porosity (eff) [vol.%]
45–85	8–10	0.2–1	13–15*

* Simper (2018), Mirwald et al. (1988). Other values: projects TH Köln, DBU-Project (2006).

Quarrying and use

The DT has good qualifications as a building stone. Cutting is easy due to the oriented feldspars. The rock could be extracted in large blocks. The location of the *Drachenfels* directly on the Rhine enabled easy transport by water (Figure 2.14.1). From the mountain, the stone only had to be transported down to the Rhine bank via so-called stone chutes and loaded on flat-bottom ships. The quarries of the DT were located directly below the summit, on the mountain slopes, in block fields of the Rüdenet, the *Drachenburgpark*, and even in the river Rhine. Depending on the quarrying location, the rock varies in its properties. During the approximately 2,000 years of mining activity, 400,000–450,000 m³ of trachyte have been extracted (Röder 1974).

As early as AD 50 to 400, the Romans quarried DT under military command and shipped it on the Rhine to Remagen and north to Xanten and Nymwegen (Röder 1974).

The real boom of the DT was caused by huge medieval building projects like the Cologne Cathedral. The Cologne "*Domfabrik*" got its own quarry "*Domkaule*" on the *Drachenfels*. There the cathedral works quarried until the first half of the 16th century (Leven 1954).

Since the end of the 17th century, the quarries of the *Drachenfels* were inactive. The stone was not suitable for the elaboration of the finer details popular in the Baroque.

During the restoration and finishing of Cologne Cathedral in the 19th century, several new approaches were made to reopen the quarries. However, at the end of the 18th century, a feeling for monument and nature conservation emerged that finally prevented the reopening of the quarries. The *Siebengebirge* was put under nature protection in 1922 (Hardenberg 1968).

During Roman times, the distribution of the DT followed the Rhine as a transport route. Memorial stones, votive stones, gravestones and building stones were made from this material.

In the Middle Ages, the material became so popular that it was distributed all over the Rhineland, up to the Netherlands. The DT was the building stone for Cologne Cathedral, Limburg and Xanten Cathedrals, the Romanesque churches in Cologne and the region as well as for public buildings and even townhouses.

DT was particularly suitable as a construction stone. Detailed decoration and sculpture are hindered by the large sanidine crystals, which can fall out already during the carving or later due to weathering. This gives the rock a pockmarked surface.

Weathering and conservation

The DT is a good and generally not weather-prone building material. After several hundred years of weathering, the DT proved in 1818 to be still in good condition at Cologne Cathedral (Scheuren 2004). However, the rock is sensitive to damaging salts. Salt weathering tests showed that DT decays almost completely within only 17–20 cycles (Luckat 1975). Therefore, rapid stone degradation occurred under the aggressive SO_2-rich atmosphere in Cologne. Weathering behavior of the DT is mainly determined by its inhomogeneous structure with large, poorly attached sanidine crystals and its pore size distribution. The most common weathering patterns are cracks, flaking, scale formation and loss of sanidine crystals (Figure 2.14.2). Often only the uppermost 1–2 cm are weathered.

Successful conservation interventions have been carried out and research still is continuing (DBU Project 2006; Plehwe-Leisen et al. 2007).

2.15   Rhenish Tuffs (Rheinische Tuffe)

Friedrich Häfner

Brief geological description

The eastern volcanic area of the Eifel Mts. covers an area of 400 km² (Bogaard & Schmincke 1990; LGB 2005). The subsurface consists of folded rocks of the lower Devonian and partly of Tertiary and Quaternary sediments of the Neuwied Basin (LGB 2005). From overall 370 volcanic structures in the Eifel Mts., about 120 are located in their eastern part. The output comprised nephelinite, leuzitite, basanite and tephrite rocks. Additionally, intermediate and phonolithic magmas formed. The tuffs of the eastern Eifel Mts. go back to phonolitic magmas producing pyroclastic flows and tephra.

Geological age

The phase of dominant volcanism existed about 20–30 Mio. years ago in the Tertiary and found its continuation in the Quaternary. All tuffs have an age of less than 1 Mio. year. The eruption of the Laacher See volcano 12,900 years ago implied the apparent end of volcanic activity in the area. However, micro earthquakes and gas emergences still can be observed. Thus, in terms of volcanology, the volcanism of the Eifel Mts. is still active.

Occurrence, quarrying and use

The tuff deposits are concentrated in the valleys of some small rivers as there are Brohl, Nette and Ahr as well as in the Laacher See area and the surroundings of the communities Kruft, Weibern, Ettringen, Kempenich, Bell and Rieden. The Romans already extracted the Rhenish Tuffs by underground mining with mechanical tools. One Roman tuff mine near Kruft (eastern Eifel Mts.) was discovered in connection with actual open cast mining. This mine has been protected by a big shelter and now is an archaeological and geological museum within the regional volcanic park. Today the deposits of Weibern and Ettringen are primarily extracted in some few open pit mines (Figure 2.15.1). The exploitation is going on using specially constructed machines that look like very big chain saws. The Rhenish Tuffs are marketed by trade names like Ettringen Tuff (*Ettringer Tuff*), Romans Tuff (*Römer-Tuff*) or Weibern Tuff (*Weiberner Tuff*). Tuffs from other deposits (*Riedener Tuff, Laacher See Tuff*) are no longer extracted.

The tuffs have a broad range of usage. Already in historic times, the tuffs have been used for masonry, sculptures, sarcophagi, aqueducts, water pipes and capitals. Some Roman towns at the Rhine River like Bonn (Vicus Bonnensis), Cologne (Colonia Agrippina), Xanten (Colonia Ulpia Traiana) and Koblenz (Ad Confluentes) used the Rhenish Tuffs to a big extent in their building activities. Today the tuffs are mainly used for restoration works, façades and sculptures.

Many historic examples of application can be still seen in the previously mentioned cities, which have a Roman background. Currently, a huge number of profane, ecclesiastical and public buildings in Germany and the neighboring countries show façades, masonry, decoration elements and sculptures of Rhenish Tuffs.

Some famous buildings in Berlin made with these tuffs are the Aquarium of the Zoological Garden (façade with reliefs), the Memorial Church (*Gedächtniskirche*) and the Charlottenburg Gate (*Charlottenburger Tor*).

Figure 2.15.1 Portal of the High Court in Koblenz (Rhine) with rich decorated ornamental stones of Weibern Tuff (2015).

Other examples are the church St. Adolfus in Düsseldorf, the county administration building and the High Court in Koblenz (Figure 2.15.1), the cathedral in Bonn, the Old Town Hall in Cologne, the cathedrals in Aachen (UNESCO World Heritage Site) and in Utrecht (NL) and the Maria Laach Abbey in the eastern Eifel Mts. (Figure 2.15.4). Rhenish Tuff is one of the German ornamental stones that is used in the whole country as well as in the neighboring countries.

Petrographical characteristics

The Rhenish Tuffs are fine-grained to coarse-grained, homogeneous tephra with a porphyric and mosaic-like structure. The amount of components differs between 76 and 97 vol.%, the visible pore volume between 3 and 24 vol.%. The colors vary from light brown, brownish-grey, greenish-grey, light olive grey to a light yellow ocher. The size of the inclusions ranges from less than 1 cm up to more than 10 cm in diameter. Besides volcanic fragments, they consist of sandstone, graywacke, slate and gneiss.

Figure 2.15.2 Thin section of Weibern Tuff with high amount of matrix.

Source: Braun 1995.

Figure 2.15.3 Quarry extracting Rhenish tuff (Weibern Tuff) in the eastern Eifel Mts.

Figure 2.15.4 Façade of the basilica of Maria Laach Abbey (13th century). Masonry and decorations are made of several sorts of Rhenish Tuffs, Kylltal Sandstone, Rhenish Basalt Lava and a limestone from Lorraine (France).

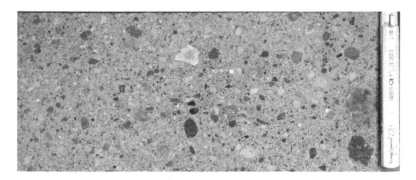

Figure 2.15.5 Sample of Rhenish Tuff (Ettringen Tuff).

Source: Photo M. Auras.

The tuffs are classified as tuffs rich on epiclasts, lapilli and ash, or as phonolithic and leucitic tuffs, respectively (Grimm 2018; Braun 1995; Schumacher & Müller 2011) (Figure 2.15.5).

Technical parameters

Weathering resistance

The Rhenish Tuffs normally have a good resistance against weathering and freezing. This is amazing in respect to the relatively high water absorption and porosity. A typical weathering phenomenon is the loss of weak components, forming holes.

Table 2.15.1 Technical Parameters of the Rhenish Tuffs

	Rhenish Tuff
Bulk density [g/cm³]	1.37–1.67
Pure density [g/cm³]	2.30–2.57
Porosity [vol.-%]	22.34–46.21
Water adsorption [wt.%]	13.21–24.76
Compressive strength [MPa]	18–36
Flexural strength [MPa]	1.4–5.1

Source: Grimm 2018; Braun 1995.

2.16   Habichtswald/Kassel Tuff

Enno Steindlberger

Brief geological description

West of Kassel, the low mountain range of the Habichtswald is typified by volcanic tuffs. This stone type was used primarily to build the monuments in the *Bergpark* Wilhelmshöhe. The trade names are Habichtswald or Kassel Tuff. During the Miocene, explosive basaltic volcanic eruptions ejected magmatic slag particles as well as xenolithic components from the underlying geological units. The primary loose deposit masses, reaching ash to lapilli size but rarely larger bomb size, too, were mostly rearranged and subsequently consolidated. The spectrum of colors and structures vary within a wide range from mostly grey to brownish or yellowish, and they can be non-layered or well layered.

Geological age and context of origin

Tertiary – Miocene; 20–7 Ma., maximum at 13–12 Ma (Wedepohl 1978).

These volcanic activities were directly connected to tectonic faults during the Alpine Orogeny. Basalts were mainly produced, but highly explosive eruptions also took place, ejecting ash and lapilli, which determine the main components of the tuff. These phreatomagmatic events can be described as interaction of magma with water-containing rocks or sediment layers. Repetitive periods of volcanic activity alternated with periods of dormant phases.

Occurrence, quarrying and use

The area south and west of Kassel is characteristically shaped by the northern Hessian low mountain range landscapes, widely built up by volcanic rock series. The tuff deposits in the Habichtswald region interlock with basaltic vents and covers. Due to its high susceptibility to erosion, the tuff is only preserved as relics in its former extent.

Both types of rocks were intensively mined in the past. The tuff was easy to cut out and therefore quarried in numerous pits. Figure 2.16.2 shows all known quarries, providing the building material for the construction of the objects in the *Bergpark* Wilhelmshöhe (see Chapter 1.5; Figure 2.16.1).

Additionally, some tuffs have been used as building material for the construction of churches and houses in the surrounding region. All

Figure 2.16.1 The artificial ruins of the *Löwenburg* Castle, built of different types of tuff.

historical quarries are nowadays dilapidated and exploited (Steindl-berger 2003). Even the protected landscape area hardly allows any new quarrying. Only in the 1980s and from 2007 onward, a quarry in the Drusel Valley has been opened for a short time (Figure 2.16.2), in each case especially for restoration purposes.

Petrographical characteristics

The tuffs are characterized by a certain variability in color and struc-ture (Figure 2.16.3). The greyish to brownish color is mainly deter-mined by the proportions of primary magmatic, glassy to aged lapilli and ash components. Because of palagonitic alteration, they may also have a yellowish color. A distinction can be made between coarser, often unlayered lapilli tuffs (2–64 mm) and fine-layered ash tuffs (< 2 mm); bigger volcanic bombs (> 64 mm) are partially embodied. Xenolithic material, as there are quartz sand grains, sandstone and basalt, derives from the underlying basement and is contained in varying proportions. Only the tuff from the Drusel Valley additionally contains typifying olivine aggregates (Figure 2.16.3 top left). The pyroclasts are black to

Figure 2.16.2 Geological overview of the volcanic rock deposits west of Kassel with the named tuff quarries that supplied the building material to the *Bergpark* Wilhelmshöhe.

Source: Steindlberger 2011.

yellowish-brown and slaggy or dense, often with a fluidal structure. Idiomorphic or corroded crystals are of basaltic nature, the most important ones are olivine, plagioclase, clinopyroxene (augite/titanaugite) and magnetite (Figure 2.16.4).

All these minerals built up the ash matrix, together with glassy and clayey particles, too. The binder consists of altered ash matrix or clay

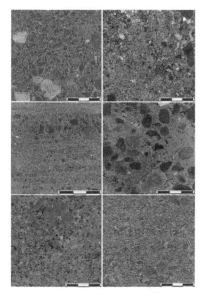

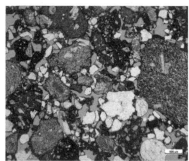

Figure 2.16.4 Thin section of a tuff from the Octagon building in the *Berg-park* showing slaggy ash grains, quartz sand (white) and a very high porosity (blue dyed).

Figure 2.16.3 Tuffstone varieties illustrating the color and structural variance. Scale: 3 cm.

mineral cement, originated by hydrothermal effects. The free interparticle space is differently porous, depending on the primarily space-filling ash content or the secondary clay mineral compaction.

Technical parameters

The properties of tuffs are strongly determined by their explosive volcanic origin. The main characteristics are high porosity and low strength. The individual values of the analyzed Habichtswald Tuffs differ in a wide range. Porosity peaks at 38.7 vol.% and mostly consists of air pores and so-called gel pores (up to 30 nm). These smallest pores are filled with water by the hygroscopic humidity, when exposed to weathering, leading to a constantly high moisture content (= water retention). Additional water can penetrate by capillary transport. The high porosity is associated with a low bulk density and a low compressive strength as well as a low flexural strength (Table 2.16.1). The notable hygric expansions are caused by amounts of smectitic clay minerals.

Table 2.16.1 Technical Parameters of Habichtswald Tuffs from Several Quarries

	Compressive Strength [MPa]	Flexural Strength [MPa]	Water Absorption (atm) [wt.%]	Porosity (eff) [vol.%]
Druseltal	34.6–37.4	0.8–1.4	10	21.6
Ahrensberg	6.9–19.8	–	15.2	35.2
Teufelsmauer	17.7	–	16.3	33.8
Essigberg	24.9	1.3–1.8	21.2	36.7
Hunrodsberg	–	–	18.9	38.7

Source: Steindlberger 2011.

Weathering resistance

The tuffs show an intense weathering, caused by a rather weak binding and a high porosity. Due to the high water absorption ability, especially inside exposed walls and socles, the tuffs are affected by frost-thaw-processes, leading to disaggregation with cracking, crumbling and sanding of the substrate. Another damage process is causally related to a high amount of clay minerals mainly within the matrix. Cyclic moistening and drying effects lead to swelling and shrinkage inside the weathered parts. Because of this, tensions finally will deform and debilitate the structure. Scaling and flaking can develop in the outer zones.

Bibliography

Auras, M. (2014) Charakterisierung von schwarzen Krusten und anderen Ober-
flächenveränderungen der Sandsteine der Porta Nigra in Trier. In *Reinigung
der Porta Nigra? Naturwissenschaftliche und restauratorische Aspekte zur
Verschwärzung und Reinigung der Sandsteine, Berichte des IfS Nr. 47*, 15–27.
Mainz: Institut für Steinkonservierung e.V.

Bachmann, G. H. & Glässer, W. (2015) Der historische Gartentisch des Refor-
mators Philipp Melanchthon in Wittenberg. *Hallesches Jahrbuch für Geowissenschaften*, 56: 122–136.

Balgstedter Chronik. 1152–1397. www.balgstaedt.de/chronik-von-1152-1397/
(accessed December 7, 2020).

Bauer-Bornemann, U. (2012) Die Restaurierung der Figurengruppe an der
Dreikönigskapelle in Naumburg. *Denkmalpflege in Sachsen-Anhalt*, 1(2012):
61–71.

Beierlein, P. R. with contribution by W. Quellmalz (1963) Geschichte der erzge-
birgischen Marmorbrüche, insbesondere des schwarzen Bruches zu Kalkgrün
bei Wildenfels. *Jahrbuch des Staatlichen Museums für Mineralogie und Geologie zu Dresden*, 1963: 163–251.

Bergmann, U. & Plehwe-Leisen, E.v. (2019) Das Recycling des römischen
Kalksteins aus Lothringen in der Kölner Bildhauerkunst des Mittelalters.
Geschichte in Köln, 66: 7–53.

Berres, F. (1996) *Gesteine des Siebengebirges*. Siegburg: Rheinlandia Verlag.

Beyer, F.M. (1996) *Geowissenschaftliche Bestandsaufnahme und Untersuchung der petrographischen und gesteinstechnologischen Eigenschaften von
mesozoischen Naturwerksteinen der Westeifel*. Diploma thesis, Johannes-
Gutenberg-Universität, Mainz.

Beyer, J. & Grimm, W.D. (1997) Der Jura-Marmor als Naturwerkstein und
Denkmalgestein. *Naturstein*, 3(97): 71–75.

Bogaard, P. v. d. & Schmincke, H.-U. (1990) Vulkanologische Karte der Osteifel
1: 50.000, Koblenz.

Braun, E. (1995) *Petrographische und petrophysikalische Untersuchungen an
Werksteinen der Osteifel*. Diploma thesis, Johannes-Gutenberg-Universität,
Mainz.

Bungert & Wirtz (2020) www.bungert-wirtz.de/Leistungen/ (accessed June 14, 2020).

Burggraff, P. (2013) Extra muros: das Zisterzienserkloster Maulbronn in der Landschaft. In *Werksteinabbau und Kulturlandschaft. Chancen und Konflikte für das Natur- und Kulturerbe. Dokumentation der Tagung am 22. und 23. März 2012 in Maulbronn*, 116–126. Bonn: Bund Heimat und Umwelt in Deutschland.

Claussen, H. & Skriver, A. (2007) *Die Klosterkirche Corvey. Band 2: Wandmalerei und Stuck aus karolingischer Zeit*. Münster: Philipp von Zabern GmbH.

Cloos, H. & Cloos, E. (1927) Die Quellkuppe des Drachenfels am Rhein. Ihre Tektonik und Bildungsweise. *Zeitschrift für Vulkanologie*, XI: 33–40.

Cüppers, H. (1990) *Die Römer in Rheinland-Pfalz*. Stuttgart: Konrad Theiss Verlag.

Dallmeier, L.M., et al. (2004) *Die Porta Praetoria in Regensburg*. Regensburg: Historischer Verein für Oberpfalz und Regensburg.

Deutsche Bundesstiftung Umwelt (DBU) (2006) *Modellhafte Entwicklung von Konservierungskonzepten für den stark umweltgeschädigten Trachyt an den Domen zu Köln und Xanten. Abschlussbericht Deutsche Bundesstiftung Umwelt Az 20105, Ref. 45*. Osnabrück: Deutsche Bundesstiftung Umwelt (DBU).

Ehlen, H.W. (ed.) (2011) *Die Rose neu erblühen lassen*. Trier: Paulinus.

Ehlers, M. (2013) Ein Bilderbuch der Geschichte und Geographie: Steinabbau in Maulbronn. In *Werksteinabbau und Kulturlandschaft. Chancen und Konflikte für das Natur- und Kulturerbe. Dokumentation der Tagung am 22. und 23. März 2012 in Maulbronn*, 127–135. Bonn: Bund Heimat und Umwelt in Deutschland.

Ehling, A. & Siedel, H. (ed.) (2011) *Bausandsteine in Deutschland, Band 2 Sachsen-Anhalt, Sachsen und Schlesien (Polen)*. Stuttgart: Schweizerbart'sche Verlagsbuchhandlung.

Elmshäuser, K., Hoffman, H.-C. & Manske, H. J. (eds.) (2002) *Das Rathaus und der Roland auf dem Marktplatz in Bremen*. Bremen: Edition Temmen.

EU Nanocathedral Project (2015–18) www.nanocathedral.eu/ (accessed December 7, 2020).

Fachbach, J. (2011) *Der Wiederaufbau der Trierer Römerbrücke 1716–1718: Ein Beispiel für den Umgang mit antiker Architektur in der frühen Neuzeit: Funde und Ausgrabungen im Bezirk Trier*. www.journals.ub.uni-heidelberg. de (accessed June 20, 2020).

Fitzner, B. & Heinrichs, K. (1992) Verwitterungszustand und Materialeigenschaften der Kalksteine des Naumburger Doms. In *Jahresberichte Steinzerfall-Steinkonservierung 2*, 1990, 23–38. Berlin: Ernst & Sohn.

Frechen, J. & Vieten, K. (1970) Petrographie der Vulkanite des Siebengebirges. *Decheniana* 122: 337–356.

Gramatzki, R. (1994) *Das Rathaus in Bremen*. Bremen: H. M. Hauschild.

Graue, B., Siegesmund, S. & Middendorf, B. (2011) Quality Assessment of Replacement Stones for the Cologne Cathedral: Mineralogical and Petrophysical Requirements. *Environmental Earth Sciences*, 63: 1799–1822.

Grimm, W.-D. (ed.) (2018) *Bildatlas wichtiger Denkmalgesteine der Bundesrepublik Deutschland* (2nd ed.). Ulm: Ebner-Verlag.

Groß-Morgen, M., director "Museum am Dom". (Personal communication, June 11, 2020).

Grunert, S. (1986) Der Sandstein der Sächsischen Schweiz. *Abhandlungen des Staatlichen Museums für Mineralogie und Geologie zu Dresden*, 34: 1–155.

Haberland, D. & Huster, H. (2007) Die Instandsetzung des Herkules-Bauwerks in Kassel-Wilhelmshöhe. *Denkmalpflege & Kulturgeschichte*, 4: 2–8.

Häfner, F. (2008) Zur Geologie der Werksteinvorkommen der Fa. Mayko. In *Mayener Basaltlava – Zeitzeuge aus den Tiefen der Vulkaneifel*, ed. J. Netz, 11–24. Mayen.

Hagdorn, H., Szulc, J., Bodzioch, A. & Morycowa, E. (1999) Riffe aus dem Muschelkalk. In *Trias – Eine ganz andere Welt*, ed. N. Hauschke & V. Wilde, 309–320. München: Verlag Dr. F. Pfeil.

Hardenberg, Th. (1968) Der Drachenfels – Seine "Conservation vermittelst Expropriation". *Rheinischen Denkmalpflege*, 4: 274–310.

Hartenstein, O. & Lange, P. (1991) Saalburger Marmor – Vorkommen, Gewinnung und Verarbeitung. *Fundgrube*, 27: 69–76, 130–37, Berlin.

Hollerung Restaurierung GmbH (2013) *Dokumentation Naturstein – und Konservierungsarbeiten Klosterkirche Maulbronn*. www.hollerung.com/uploads/tx_hollerung/fotodokumentation-maulbronn-2013.pdf (accessed December 7, 2020).

Holzhauser, P. (2011) Bestimmung der Scherfestigkeit an veränderlich festen Gesteinen in Hinblick auf Hangbewegungsphänomene. In *Münchner Geowissenschaftliche Abhandlungen* (Reihe B: Bd. 17). München: Verlag Dr. F. Pfeil.

Hoppe, W. (1939) *Vorkommen und Beschaffenheit der Werk- und Dekorationssteine in Thüringen*. Berlin: Union Deutsche Verlagsgesellschaft Roth & Co.

Jelschewski, D. (2020) TU München. (Personal communication, July 1–3, 2020).

Katzschmann, L., Aselmeyer, G. & Auras, M. (2006) Natursteinkataster Thüringen. In *Berichte des IfS 23*, 1–196. Mainz: Institut für Steinkonservierung e.V.

Katzschmann, L. & Lepper, J. Werksteine der Germanischen Trias. In *Trias – Eine ganz andere Welt*, ed. N. Hauschke & V. Wilde – unpublished manuscript of 2nd edition, 2020 (first edition 1999, München: Verlag Dr. F. Pfeil).

Keupp, H. (1977) Ultrafacies und Genese der Solnhofener Plattenkalke (Oberer Malm, Südliche Frankenalb). *Abhandlungen der Naturhistorischen Gesellschaft*, 37: 1–128.

Kirnbauer, T. (2008) Nassau Marble or Lahn Marble (Lahnmarmor) – A Famous Devonian Dimension Stone from Germany. *Schriftenreihe der Deutschen Gesellschaft für Geowissenschaften*, 59: 187–218.

Klaua, D. (1964) *Sedimentpetrographische Untersuchungen zur Verwitterung der Rätsandsteine Thüringens*. PhD dissertation, Hochschule für Architektur und Bauwesen, Weimar.

Koch, R. (2007) Sedimentological and Petrophysical Characteristics of Solnhofen Monument Stones – Lithographic Limestone: A Key to Diagenesis and

Fossil Preservation. *Neues Jahrbuch für Geologie und Paläontologie*, 245(1): 103–115.

Koch, R., Sobott, R. & Lorenz, H.G. (1999) Der Schaumkalk (Trias, Unterer Muschelkalk) am Naumburger Dom als Baustein: Einfluß von Fazies und Diagenese auf die Gesteinsqualität. In *Trias – Eine ganz andere Welt*, ed. N. Hauschke & V. Wilde, 449–471. München: Verlag Dr. F. Pfeil.

Kott, R. (1989) Fazies und Geochemie des Treuchtlinger Marmors (Unter- und Mittel-Kimmeridge, Südliche Frankenalb). *Berliner geowissenschaftliche Abhandlungen / A.*, 111. Berlin.

Kraus, K. & Jasmund, K. (1981) Verwitterungsvorgänge an Bausteinen des Kölner Doms. *Kölner Domblatt*, 46: 175–190.

Kuster-Wendenburg, E. (2002) *Der Bremer Stein und die Weserrenaissance: Bremer Geo-Touren*, ed. G. Wefer, Heft 1. Bremen: marum/rcom bibliothek. www.marum.de/Binaries/Binary_3102/Geo-Touren-1.pdf (accessed December 7th, 2020)

Kutschke, D. (2000) Steinbrüche und Steinbrecher in der Sächsischen Schweiz. *Schriftenreihe des Stadtmuseums Pirna*, 11: 1–200. Pirna.

Leopold, G. & Schubert, E. (1972) *Die frühromanischen Vorgängerbauten des Naumburger Doms*. Berlin: Akademie-Verlag.

Lepper, J. & Ehling, A. (2018a) Wesersandstein. In *Bausandsteine in Deutschland, Band 3A Niedersachsen*, ed. A. Ehling & J. Lepper, 65–107. Stuttgart: Schweizerbart'sche Verlagsbuchhandlung.

Lepper, J. & Ehling, A. (2018b) Wealden-Sandsteine. Obernkirchener-, Rehburger-, Deister-, Süntel-, Nesselberg- und Osterwald-Sandstein. In *Bausandsteine in Deutschland, Band 3A Niedersachsen*, ed. A. Ehling & J. Lepper, 230–265. Stuttgart: Schweizerbart'sche Verlagsbuchhandlung.

Lepper, J., Jung, G. & Seibertz, E. (2018) Baulandschaft und Bausteine zwischen Oker und Aller (Östliches Braunschweiger Land). *Jahresberichte und Mitteilungen des Oberrheinischen Geologischen Vereins*, N.F., 100: 169–209.

Leven, H. (1954) Beiträge zur Geschichte der Steinbruch- und Steinmetzbetriebe am Siebengebirge. *Bonner Geschichtsblätter*, 8: 135–165.

LGB – Landesamt für Geologie und Bergbau (ed.) (2005) *Geologie von Rheinland-Pfalz*. Stuttgart: Schweizerbart.

Liebfrauenkirche (2020) https://de.wikipedia.org/wiki/Liebfrauenkirche_(Trier) (accessed June 8, 2020).

Lobbedey, U. (2007) *Corvey – Church, Former Monastery, Now Schloss and Grounds*. München & Berlin: Deutscher Kunstverlag.

Löhr, H. (2015) Brandspuren an der Porta Nigra in Trier. *Funde und Ausgrabungen im Bezirk Trier*, 36–50, www.journals.ub.uni-heidelberg.de (accessed June 20, 2020).

Luckat, S. (1975) Die Einwirkung von Luftverunreinigungen auf die Bausubstanz des Kölner Doms III. *Kölner Domblatt*, 40: 75–188.

Mangartz, F. (2000) Römerzeitlicher Abbau von Basaltlava in der Osteifel Ein bedeutender Wirtschaftszweig der Nordwestprovinzen. In *Steinbruch und*

Bergwerk: Denkmäler römischer Technikgeschichte zwischen Eifel und Rhein, Vulkanpark Forschungen 2, 6–16. Mainz: Verlag des römisch-germanischen Zentralmuseums.

Mirwald, P., Kraus, K. & Wolff, A. (1988) Stone Deterioration on the Cathedral of Cologne. *Durability of Building Materials*, 5: 549–570.

Nancy-Metz (2020) http://www4.ac-nancy-metz.fr/base-geol/fiche.php?dossier= 199&p=3descrip (accessed June 8, 2020).

Neumann, H.-H. (1994) *Aufbau, Ausbildung und Verbreitung schwarzer Gipskrusten, dünner schwarzer Schichten und Schalen sowie damit zusammenhängender Gefügeschäden an Bauwerken aus Naturstein.* PhD dissertation, Universität Hamburg, Schriftenreihe Angewandte Analytik 24.

Nipperdey, Th. (1968) Nationalidee und Nationaldenkmal in Deutschland im 19. Jahrhundert. *Historische Zeitschrift*, 206(3): 529–585.

Ostermann, P. (2001) *Denkmaltopographie Bundesrepublik Deutschland. Kulturdenkmäler in Rheinland-Pfalz: Band 17.1: Stadt Trier, Altstadt*, ed. Landesamt für Denkmalpflege. Worms: Wernersche Verlagsgesellschaft.

Paulinus (2020) www.paulinus.de/archiv/archiv/9918/bistuma2.htm (accessed June 8, 2020).

Plehwe-Leisen, E. von (2004) Die Gesteine des Domes. In *Steine für den Kölner Dom*, ed. B. Schock-Werner & R. Lauer, 78–97. Köln: Verlag Kölner Dom.

Plehwe-Leisen, E. von (2005) Bewitterungsexperimente am Kölner Dom und ihre Bedeutung für die praktische Denkmalpflege. *Zeitschrift der Deutschen Gesellschaft für Geowissenschaften*, 156(1): 159–166.

Plehwe-Leisen, E. von (2007a) Die mittelalterliche Skulptur des Kölner Doms unter dem Mikroskop. *Kölner Domblatt*, 72: 149–160.

Plehwe-Leisen, E. von (2007b) In Peter Mirwalds Fußstapfen – Aspekte der Steinkonservierungsforschung am Kölner Dom. In *Naturwissenschaft und Denkmalpflege*, ed. A. Diekamp, 235–243. Innsbruck: Innsbruck University Press.

Plehwe-Leisen, E. von & Leisen, H. (2015) Barrois-Oolith am Dom zu Köln. In *Barrois-Oolithe*, ed. G. Lehrberger & E. von Plehwe-Leisen, 194–201. München: Verlag Dr. F. Pfeil.

Plehwe-Leisen, E. von & Leisen, H. (2019) Geowissenschaftliche Untersuchungen zur Verwendung von Naturstein in der mittelalterlichen Bildhauerkunst in Köln. *ZKK Zeitschrift für Kunsttechnologie und Konservierung*, 33(1): 7–32.

Plehwe-Leisen E. von, Leisen, H. & Wendler, E. (2007) Der Drachenfels-Trachyt – ein wichtiges Denkmalgestein des Mittelalters – Untersuchungen zur Konservierung. *Zeitschrift der Deutschen Gesellschaft für Geowissenschaften*, 158(4): 985–995.

Poschlod, K., Sutterer, V., Wamsler, S. & Woznik, E. (2017) Erkundung und Untersuchung von Regensburger Grünsandstein. *Umwelt Spezial*: 1–74. Augsburg: Bayerisches Landesamt für Umwelt.

Poschlod, K. & Wamsler, S. (2009) Green Sandstone for the Stone Bridge in Regensburg. In *Proc. of the 6th European Congress on Regional Geoscientific*

Cartography and Information Systems (Vol. II, 206–207). Munich: Bayerisches Landesamt für Umwelt.

Przybyla, T., Pfänder, J.A., Münker, C., et al. (2018) High-resolution ⁴⁰Ar/³⁹ Ar Geochronology of Volcanic Rocks from the Siebengebirge (Central Germany) – Implications for Eruption Timescales and Petrogenetic Evolution of Intraplate Volcanic Fields. *International Journal of Earth Sciences (Geologische Rundschau)*, 107: 1465–1484.

Raddatz-Antusch, M. (2019) Geologie und Paläontologie der unterkreidezeitlichen Sandsteine des Bückebergs bei Obernkirchen (Niedersachsen). *Naturhistorica, Berichte der Naturhistorischen Gesellschaft Hannover*, 161: 7–98, Hannover.

Ringbeck, B. et al. (2012) *Executive Summary and Nomination for Inscription on the UNESCO World Cultural and Natural Heritage List: The Carolingian Westwork and the Civitas Corvey.* http://whc.unesco.org/en/list/1447/documents (accessed December 7, 2020)

Ritter-Höll, A. 2005. *Werksteinqualitäten im Jura-Kalkstein anhand des Steinbruchs Max Balz, Pappenheim.* PhD dissertation, Universität Erlangen-Nürnberg, München, 1–151.

Röder, J. (1974) Römische Steinbruchtätigkeit am Drachenfels. *Bonner Jahrbücher*, 74: 509–542.

Ruppienė, V. (2021) Pavements and revetments in the audience hall (Basilika) and its vestibule of the late-antique imperial palace in Trier (Germany). In *Stone and Splendor. Interior decorations in late-antique palaces and villas,* Proceedings of a workshop, Trier, 25–26 April 2019, ed. V. Ruppienė. *Forschungen zu spätrömischen Residenzen* 1, 37–53. Wiesbaden: Harrassowitz.

Sander, H. (1981) *Das Herkules-Bauwerk in Kassel-Wilhelmshöhe.* Kassel: Thiele & Schwarz.

Savchenok, A. (2009) *Mineralogy of Building Stones at Architectonical Objects in St. Petersburg – Composition, Features and Weathering Behaviour in a City Atmosphere (in Russian).* PhD dissertation, St. Petersburg State University, Department of Mineralogy, St. Petersburg.

Schaaff, H. (2000) Antike Tuffbergwerke in der Pellenz. In *Steinbruch und Bergwerk: Denkmäler römischer Technikgeschichte zwischen Eifel und Rhein,* Vulkanpark-Forschungen, 2, 17–30. Mainz: Verlag des römisch-germanischen Zentralmuseums.

Scheuren, E. (2004) Kölner Dom und Drachenfels. In *Steine für den Kölner Dom,* ed. B. Lambert & T. Weber, 22–45. Köln: Verlag Kölner Dom.

Schirrmeister, G. (2006) Weltkulturerbe Museumsinsel. In *Naturwerksteine für Architektur- und Baugeschichte von Berlin,* ed. J. H. Schroeder, 121–130. Berlin: Selbstverlag Geowissenschaftler in Berlin und Brandenburg e. V.

Schirrmeister, G. (2009) Tests on Selected Stonework Materials. In *The Neues Museum Berlin. Conserving, restoring, rebuilding within the World Heritage,* ed. Staatliche Museen zu Berlin – Stiftung Preußischer Kulturbesitz, Bundesamt für Bauwesen und Raumordnung, Landesdenkmalamt Berlin, 149–150. Leipzig: Seemann.

Schirrmeister, G. (2013) Natursteine im Neuen Museum Berlin. In *Konservierung, Restaurierung und Ergänzung im Neuen Museum Berlin. Zwischen ursprünglichen Intentionen und neuen Entwürfen*, ed. Verband der Restauratoren e. V., 204–211. München: Siegl.

Schubert, E. (1997) *Der Naumburger Dom*. Halle: Stekovics.

Schumacher, K.-H. & Müller, W. (2011) *Steinreiche Eifel. Herkunft, Gewinnung und Verwendung der Eifelgesteine*. Mendig, Koblenz: Görres Druckerei u. Verlag.

Schumacher, Th. (1993) Großbaustelle Kölner Dom. *Studien zum Kölner Dom 4*. Köln: Verlag Kölner Dom.

Schumacher, Th. (2004) Steine für den Dom. In *Steine für den Kölner Dom*, ed. B. Schock-Werner & R. Lauer, 46–77. Köln: Verlag Kölner Dom.

Schwarz, Ch. (2014) *Die Geschichte der geologischen Erforschung des Siebengebirges.* https://virtuellesbrueckenhofmuseum.de/vmuseum/historie_data/dokument/Geschichte_geologischen_Erforschung_des_Siebengebirges. pdf (accessed June 17, 2020).

Seidel, G. & Steiner, W. (1988) Baustein und Bauwerk in Weimar. *Tradition und Gegenwart – Weimarer Schriften*, 31: 1–96.

Siedel, H. (2010) Alveolar Weathering of Cretaceous Building Sandstones on Monuments in Saxony, Germany. *Geological Society of London Special Publication*, 333: 11–23.

Siedel, H. (2013) Recording Natural Stones on Façades as a Tool to Assess Their Utilization and Functional Aspects Over Time. *Quarterly Journal of Engineering Geology and Hydrogeology*, 46: 439–448.

Siedel, H., Götze, J., Kleeberg, K. & Palme, G. (2011) Bausandsteine in Sachsen. In *Bausandsteine in Deutschland* (Vol. 2), ed. A. Ehling & H. Siedel, 161–270. Stuttgart: Schweizerbart'sche Verlagsbuchhandlung.

Simper, M. (2018) Gestein Nr. 042 Drachenfels-Trachyt. In *Bildatlas wichtiger Denkmalgesteine der Bundesrepublik Deutschland, Teil II* (2nd ed.), ed. W.-D. Grimm & R. Koch, 94–95. Ulm: Ebner Verlag.

Singewald, C. (1992) *Naturwerkstein – Exploration und Gewinnung*. Mayen, Köln: Steintechnisches Institut.

Sobott, R. (2009) Ultraschallmessungen an Werksteinen des Naumburger und Merseburger Doms. In *Stein – Zerfall und Konservierung*, ed. S. Siegesmund, M. Auras & R. Snethlage, 150–154. Leipzig: Edition Leipzig.

Sobott, R. (2013) Schaumkalk – der Werkstein des Naumburger Meisters. In *Macht. Glanz. Glaube. Auf dem Weg zum Welterbe. Eine Zeitreise in die hochmittelalterliche Herrschaftslandschaft um Naumburg*, ed. Förderverein Welterbe an Saale und Unstrut, 119–123. Halle: Stekovics.

Steindlberger, E. (2003) Vulkanische Gesteine aus Hessen und ihre Eigenschaften als Naturwerksteine. In *Geologische Abhandlungen Hessen 110*, 1–167. Wiesbaden: Hessisches Landesamt für Naturschutz, Umwelt und Geologie.

Steindlberger, E. (2011) Tuffstein als Baumaterial im Bergpark Wilhelmshöhe. In *Das Herkulesbauwerk im Bergpark Wilhelmshöhe, Berichte zur Restaurierung. Arbeitshefte des Landesamtes für Denkmalpflege Hessen 18*, 53–64. Stuttgart: Theiss Verlag.

Steindlberger, E. (2020a) Konservierung hessischer Tuffe und Schalsteine unter Langzeitbeobachtung. In *Konservierungsstrategien für Rhyolithtuff-mauerwerk, Berichte des IfS Nr. 60*, 51–70. Mainz: Institut für Steinkonservierung e.V.

Steindlberger, E. (2020b) Conservation of Northern Hessian Tuffstones. A Building Material Used for the Monuments in the Bergpark Wilhelmshöhe. In *Monument Future: Decay and Conservation of Stone. Proceedings of the 4th International Congress on the Deterioration and Conservation of Stone*, ed. S. Siegesmund & B. Middendorf, 865–870. Halle: Mitteldeutscher Verlag.

UNESCO WHC (1999) *Museumsinsel (Museum Island), Berlin.* https://whc.unesco.org/en/list/896

UNESCO WHC (2004) *Town Hall and Roland on the Marketplace of Bremen.* https://whc.unesco.org/en/list/1087/

Völker, M. (1999) *Werksteine des Muschelkalk in der Region Trier und ihre Beziehung zu römischen Baudenkmälern.* Diploma thesis, Universität Würzburg.

Wagner, H.W. (1991) Der Dolomitstein der Trier-Bitburger Mulde als Natur-werkstein-Rohstoff. *Mainzer geowissenschaftliche Mitteilungen*, 20: 85–90.

Wagner, H.W. (2012) Liebfrauen-Basilika in Trier: Die Rose blüht. *Naturstein*, 4: 50–51.

Wagner, H.W. (2020) *Rohstoffgeologie des Dachschiefers – Petrographie, Materialprüfung, Normung, Lagerstätten, aktuelle und historische Verwendung, Markt, Gewinnung, Global Heritage Stone Resource (GHSR).* Submitted habilitation thesis, not yet published, Universität Trier, 1–570.

Wagner, H.W., Kremb-Wagner, F., Koziol, M. & Negendank, J.F.W. (2012) *Trier und Umgebung (Sammlung geologischer Führer)* (Bd. 60). Stuttgart: Borntraeger Verlag.

Wagner, H.W. & Wechsler, G. (2009) Liebfrauenkirche in Trier: Die Rose soll neu erblühen. *Naturstein*, 4: 2–4.

Weber, J. & Lepper, J. (2002) Depositional Environment and Diagenesis as Controlling Factors for Petro-physical Properties and Weathering Resistance of Siliciclastic Dimension Stones: Integrative Case Study on the "Wesersandstein" (Northern Germany, Middle Buntsandstein). *Geological Society of London Special Publications*, 205: 103–114.

Weber, W. (2003) Archäologische Zeugnisse aus der Spätantike und dem frühen Mittelalter zur Geschichte der Kirche im Bistum Trier (3.–10. Jahrhundert n. Chr.). *Veröffentlichungen des Bistumsarchivs Trier*, 38: 407–541.

Weber, W. (2017) Die Trierer Domgrabung. Die Ausgrabungen in der Kurie von der Leyen und der Liebfrauenstraße (Südwest-Bereich). Teil 2. Die Befunde. In *Kataloge und Schriften des Bischöflichen Dom- und Diözesanmuseums Trier, Band 7.* Trier: Bischöfliches Dom- und Diözesanmusem.

Wedepohl, K.H. (1978) Der tertiäre basaltische Vulkanismus nördlich des Vogelsberges. *Aufschluß, Sonderband*, 28: 156–167. Heidelberg.

Wehry, M. (2005) *Die Bausandsteine des Buntsandsteins bei Nebra, Sachsen-Anhalt.* Diploma thesis, Freie Universität, Berlin.

Werner, W., Wittenbrink, J., Bock, H. & Kimmig, B. (2013) *Naturwerksteine aus Baden-Württemberg – Vorkommen, Beschaffenheit und Nutzung.* Freiburg: Landesamt für Geologie, Rohstoffe und Bergbau.

Wilmsen, M. & Niebuhr, B. (2014) The Rosetted Trace Fossil Dactyloidites Ottoi (Geinitz, 1849) from the Cenomanian (Upper Cretaceous) of Saxony and Bavaria (Germany): Ichno-taxonomic Remarks and Palaeoenvironmental Implications. *Paläontologische Zeitschrift*, 88: 123–138.

Wolff, A. (1972) Die Gefährdung des Kölner Doms. *Kölner Domblatt*, 35: 7–28.

Wolff, A. (1974) *Der Kölner Dom.* Simbach: Verlag Müller und Schindler.

Wolff, A. (2001) Zwei Erzbischöfe aus einer Säule. *Kölner Domblatt*, 66: 277–292.

Wolff, A. (2004) Stein und Bau. In *Steine für den Kölner Dom*, ed. B. Schock-Werner & R. Lauer, 8–21. Köln: Verlag Kölner Dom.

Zentrum der Antike (2020) www.zentrum-der-antike.de/en.html (accessed June 8, 2020).

Zink, J. (1980) Die Baugeschichte des Trierer Domes von den Anfängen im 4. Jahrhundert bis zur letzten Restaurierung. In *Jahrbuch des Rheinischen Vereins für Denkmalpflege und Landschaftsschutz, 1978/79*, 17–111. Neuss: Verlag Gesellschaft für Buchdruckerei AG.

Index

Milton Keynes UK
Ingram Content Group UK Ltd.
UKHW020309071024
449327UK00009B/192